# 13th International Conference on Advances in Mass Data Analysis of Images and Signals in Medicine, Biotechnology, Chemistry and Food Industry (MDA 2018)



New York, New York, USA
July 2018

**Editors:**

**Petra Perner**

ISBN: 978-1-5108-9520-1

**Printed from e-media with permission by:**

Curran Associates, Inc.
57 Morehouse Lane
Red Hook, NY 12571

**Some format issues inherent in the e-media version may also appear in this print version.**



ibai-publishing
Arno-Nitzsche-Str. 45
04277 Leipzig, Germany

info@ibai-publishing.org

**Additional copies of this publication are available from:**

Curran Associates, Inc.
57 Morehouse Lane
Red Hook, NY 12571 USA
Phone: 845-758-0400
Fax: 845-758-2633
Email: curran@proceedings.com
Web: www.proceedings.com

# Table of Contents

Petra Perner (Ed.)

# Advances in Mass Data Analysis of Images and Signals

## with Applications in Medicine, r/g/b Biotechnology, Food Industries, Dietetics, Biometry and Security, Agriculture, Drug Discover, and System Biology

13th International Conference, MDA 2018
New York, USA, July 2018

**Proceedings**

ibai Publishing
www.ibai-publishing.org

Petra Perner (Ed.)

# Advances in Mass Data Analysis of Images and Signals

## with Applications in Medicine, r/g/b Biotechnology, Food Industries, Dietetics, Biometry and Security, Agriculture, Drug Discover, and System Biology

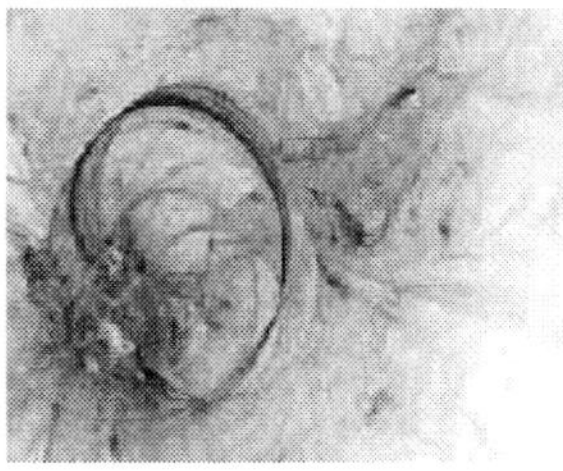

13th International Conference, MDA 2018
New York, USA, July 2018

**Proceedings**

ibai - publishing

# Preface

The 13th International Conference on Mass Data Analysis of Signals and Images with Applications in Medicine, r/g/b Biotechnology, Food Industries and Dietetics, Biometry and Security, and Agriculture MDA 2018 held in July in Ne Work, USA showed once more that the event is a must for all specialists from research and industry alike who like to stay informed about hot new topics in mass data analysis of signals and images. This year, papers maily for one general topics were presented. The analysis of CT-, Roentgen- and MRI-images has been widely studied by the presenters for different medical applications.

A tutorial on Intelligent Image Interpretation has been given in connection with the conference. It was very inspiring and taken by a large audience.
Selected revised papers of this conference as well as contributed papers will be published in the July 2018 issue of the International Journal Transaction on Mass Data Analysis of Signals and Images.

We are happy to see that new automatic systems for the analysis of signals and images have been presented at MDA 2018 and that a lot of researchers followed our mission to bring new signal and image analysis methods into real applications that introduce a new quality level to various real life applications.

The next conference will be held in 2019 in New York, USA again (www.mda-signals.de) under the auspices of the World Congress Frontiers on Intelligent Data Analysis DSA 2019 (www.worldcongressdsa.com).
We would like to invite you to contribute to this conference. Please come and join us.

July 2018

Prof. Dr. Petra Perner
Chair of the Conference MDA

The German National Library listed this publication in the German National Bibliography.
Detailed bibliographical data can be downloaded from http://dnb.ddb.de.

ibai-publishing
Prof. Dr. Petra Perner

Arno-Nitzsche-Str. 45,

04277 Leipzig, Germany
E-mail: info@ibai-publishing.org
http://www.ibai-publishing.org

## International Conference on Mass Data Analysis with Applications in Medicine, r/g/b Biotechnology, Food Industries and Dietetics, Biometry and Securitry, and Agriculture MDA 2014

### Chair

| | |
|---|---|
| Petra Perner | IBaI Leipzig, Germany |

### Committee

| | |
|---|---|
| Walter Arnold | Fraunhofer-Institute of non-destructive Testing, Germany |
| Valentin Brimkov | Buffalo State College, USA |
| Hans du Buf | University of Algarve, Portugal |
| Calin Ciufudean | "Stefan cel Mare" University, Romania |
| Kamil Dimililer | Near East University, Cyprus |
| George C. Giakos | University of Akron, USA |
| Giulio Iannello | University Campus Bio-Medico of Rome, Italy |
| Ovidio Salvetti | CNR-ISTI, Italy |
| Rainer Schmidt | University of Rostock, Germany |
| Paolo Soda | University Campus Bio-Medico of Rome, Italy |
| Thang Viet | Pham Intelligent Systems Lab Amsterdam (ISLA), The Netherlands |
| Patrick Wang | Northeastern University, USA |

# Principal Component Analysis based Multimodal Medical Image Fusion of MR and CT using Wavelet Transform

Yosef Hasan Jbara

Department of Computer Science, College of Engineering and Information Technology, Buraydha Private Colleges, Al Qassim, Kingdom of Saudi Arabia

yosef.hasan@bpc.edu.sa

**Abstract.** Image fusion of multi-model bio-medical images assists the general medical practitioner in diagnostics and therapeutic planning of the patient. The process of image fusion merges different images to make into one image so that the resultant image poses all the complementary information of the original images. The image fusion provides the complementary anatomic details of an organ, from the different modality images, during diagnostics and therapeutic planning of the patient. This research paper presents a novel and effective fusion method for medical images, by using the principal component analysis in the wavelet domain. This approach enhances both spectral and spatial quality of the fused image. The PCA technique, KL transform procedure is applied to reduce the noise and dimension of both MR and CT images, with improved geometrical resolution, which generates several principal component images. Initially, the first principal component image of source images is decomposed into different frequency bands using discrete wavelet transform. Then, the average fusion rule is applied to the coefficients of the low frequency band, and gradient weighted maximum fusion rule is applied to the coefficients of the high frequency band. Further by using inverse wavelet transform and inverse KL transform, the fused image is produced. In addition, the fused image was tested quantitatively with spatial and spectral quality measures. The spatial quality assessment is accessed by the parameters like SD, ACC, and RMSE. Similarly, the spatial quality assessment is accessed by the parameters like PSNR, AG and the entropy. The investigation shows that the presented algorithm, PCA-based wavelet technique for image fusion has the best measure concerning spectral and spatial qualities.

**Keywords:** Image Fusion, Karhunen-Loeve Transform, Multi resolution analysis, Discrete Wavelet Transform, Gradient operator, Quality Assessment.

## 1 Introduction

Since last decades, medical images acquired from recent technologies like CT, MR, MRA, PET, and SPECT have an extraordinary potential in modern medical practices on clinical diagnosis, surgery, and radiotherapy [1]. However, all these images vary by giving complementary information for the same organ. In particular, in anatomic radiology, CT image is suitably best on finding detail about thicker tissue structured organs whereas MR Image provides the best information on soft tissue organs without any information about bones [2]. In general, it is worthwhile to the medical practitioner if all the complementary information is available in a single image. The technique 'fusion' achieves the ability to maintain complimentary information in a single image. The image fusion is a promising and a very challenging task in the recent research scenario. In radiology, the impact of multispectral and multimodal image fusion image [3] fascinates ample responsiveness.

Image fusion is a method of merging different images of the similar object to get higher spectral information and sharper edges than source images by extracting the redundant and complementary information from the individual images. For fusion, it is necessary to have images either from different modalities or the same modality with different doses. At present fusion methods, pixel averaging, weighted pixel maximum, several techniques based on pyramid decomposition and wavelet decomposition fusion, are available [4].

The image fusion method based on pixel averaging technique [5] not desirable because it introduces a contrast reduction in fused image. However, the contrast reduction was eliminated in an alternate technique of weighted averaging fusion, in which the fused image contains the weighted average of intensity values. The difficulty in the alternate method is the selection of the threshold for the estimation. It is another milestone in the image fusion research is based on the pyramid and wavelet decomposition of images [6] which provide multi-resolution levels of sub-images. Researchers revealed that performing fusion in transform domain rather than the spatial domain, suits the HVS nature of human vision [7]. In pyramid transform based image fusion the coefficients of the pyramid structure of the source image, are modified according to the fusion rule. The pyramid based fused images having blocking effects and undesired edges. The next type of fusion techniques based on wavelet transform [8] does perform better on spatially and spectrally, than the pyramid based fusion, due to the nature of DWT's non-redundant image representation capability. Wavelet based image fusion methods are widely used because of its good SNR and zero blocking artifacts.

In all these methods the fusion rule is applied for all sub-images as uniform. It is not suitable and acceptable for using the same fusion rule for all sub-images because the physical meaning of the coefficients belongs to different frequency bands are different. In this research, the fusion rule is applied separately for LL and other LH, HL and HH sub-images. Pixel averaging fusion rule was applied to the coefficients belongs to low-frequency sub-images and gradient based maximum selection scheme was applied to higher frequency band sub-images. The experimental results show that both spectral and spatial quality of fused images of our scheme of fusion technique is impressive [9].

The remaining content of this research paper is given in six sections. Section2 explained KL transform techniques, and section 3 has the details of the existing wavelet-based image fusion technique. Section 4 discussed our novel fusion technique, Section 5 presents the evaluation techniques of both spatial and spectral contents of fused images, section 6 gives the experimental results with analysis, and finally, Section 7 concludes our fusion technique.

## 2 Karhunen-Loeve Transform

The KLT is a method of principal component analysis in image data. It is a unitary rotation transformation technique which aligns a data set into a sequence of the uncorrelated coefficient with the different eigen values. It preserves the essential information content of a multi-spectral (satellite images) or multi-model (bio-medical images) with a reduced number of dimensions [10]. Data Reduction is possible with the rotation transformation ability of KL transform. The orthonormal eigen vectors of autocorrelation matrix Rx of an image X (i, j) gives the basis vectors for the KL transform. The KL transform decomposes an image into principal component images called basis image. Each eigen vector produces one basis image from its corresponding eigen value. In turn, an image covariance matrix determines the choice of the principal component image [11].

The following tasks are performed to compute the KL transform for a given image of size N x N

Calculate the mean vector, which is the ratio of the sum of all intensity values of the image and to the number of pixels:

$$\mu = \frac{1}{N}\sum_{i=1}^{N} x_i$$

Derive covariance matrix using the following expression.

$$C = \frac{1}{N}\sum_{i=1}^{N} (x_i - \mu)\,(x_i - \mu)^T$$

Find the eigen values of the covariance matrix, and its corresponding eigenvectors. The order of the eigen values is $\lambda_1 > \lambda_2 > \lambda_3 \ldots\ldots\ldots\ldots > \lambda_n$.

Use the eigenvectors to form a transformation matrix 'A', whose first row is the eigenvector corresponds to the highest eigen value '$\lambda_1$' and the last row is the eigenvector corresponds to the lowest eigen value '$\lambda_n$'.

Finally, construct the transformed image using the given transform matrix.

$$y_i = A\,(x_i - \mu)$$

The covariance matrix of transformed image Y is similar to the diagonal matrix whose diagonal values are eigen values. The transformed image is less correlated than the original image.

$$C_y = \begin{bmatrix} \lambda_1 & \cdots & 0 \\ \vdots & \ddots & \vdots \\ 0 & \cdots & \lambda_n \end{bmatrix}$$

The following equation reconstructs the original image from the transformed image.

$$\hat{x} = A^T y + \mu$$

## 3. Wavelet Transform Based Image Fusion

### 3.1. Continuous Wavelet Transform

A small wave, which is used to transform a signal into sequences of orthonormal basis signal, is called mother wavelet Ψ (t). It offers a technique for investigating signals, in time-frequency scale. It has the capability to approximate the real world signal with sharp discontinuities [12]. Any signal can be approximated by combining these basis function, which are the translated and scaled version of mother wavelet Ψ (t). Translated and scaled version of mother wavelet is called as child wavelet. The mathematical equation for the child wavelet is given in the equation in which 'a' represents the scaling factor and 'b' represents the translation factor [13].

$$\psi_{a,b}(t) = \frac{1}{\sqrt{a}}\,\psi\left(\frac{t-b}{a}\right)$$

An inner product between child wavelet and function *f(t)* will give the transformed coefficient '$w_{a,b}$' which is related to a function f(t), can be achieved by the following equation.

$$w_{a,b} = \langle \psi_{a,b}, f(t) \rangle = \int_{-\infty}^{\infty} \psi_{a,b}\; f(t)\, dt$$

*f(t)* can be reconstruct from $w_{a,b}$ by double integration of the product of wavelet coefficient and mother wavelets.

$$f(t) = \frac{1}{C_\psi} \int_{-\infty}^{\infty} \int_{-\infty}^{\infty} w_{a,b}\, \psi_{a,b}(t) \frac{dadb}{a^2}$$

Wavelet transform represents a signal in time-frequency scale. This ability leads to the multi-resolution analysis of a real-time signal by avoiding the limitation in bands according to Heisenberg uncertainty principle. At higher frequencies, wavelet technique offers a good time resolution and a poor frequency resolution. However, it gives the inverse response at lower frequencies [14].

### 3.2 Discrete Wavelet Transform

The analysis and synthesis implementation of discrete wavelet transform afford necessary data calculating repetitive operations. It decreases the time for computation satisfactorily and very simple to execute. It can examine and investigate an image in multi-resolution at a different frequency band. It decomposes the image into a set of detail and approximation coefficient [15].

The forward equation for the computation of wavelet coefficient is given by

$$w_{m,n} = \langle\, f(t), \psi_{a,b} \rangle = a_0^{m/2} \int (f(t)\, \psi(a_0^m\,(t) - nb_0)dt$$

The synthesis equation to rebuild the signal *x(n)*is given by

$$x(n) = \sum_m \sum_n w_{m,n}\, \psi_{a,b}(n)$$

Wavelet split a signal into a series orthogonal functions of scaled and the translated version of a basis function, mother wavelet. In the analog implementation, the scaling and translation operations are analogous to the high pass and a low pass filter. In digital counterpart, they are a moving average and moving difference respectively [16].

Fig. 1 displays the 2D DWT analysis of an MR brain image, taken from our image sets. Fig. 1 (a), (b), (c) and (d) shows the original image, sub-bands after applying DWT, 2 level wavelet decomposition, and 3 level decomposition respectively.

The 2D-DWT technique decomposes the image into LL, LH, HL, and HH sub-frequency bands, as shown in Fig.1.b. Two level DWT gives an approximated image, the horizontal edged image, the vertical edged image and the diagonal edged image. As shown in Fig. 1. (c), Two level DWT divides the original image into four sub-images, which are shown in four quadrants. The sub-image LL, in the upper left quadrant, is an approximated version of the original image, which comprises the low-resolution coefficients of the original image. The upper right quadrant image contains the coefficients belongs to horizontal edges. The lower left quadrant contains the coefficients of vertical edges. Finally, the lower right quadrant contains the coefficients of the diagonal edges.

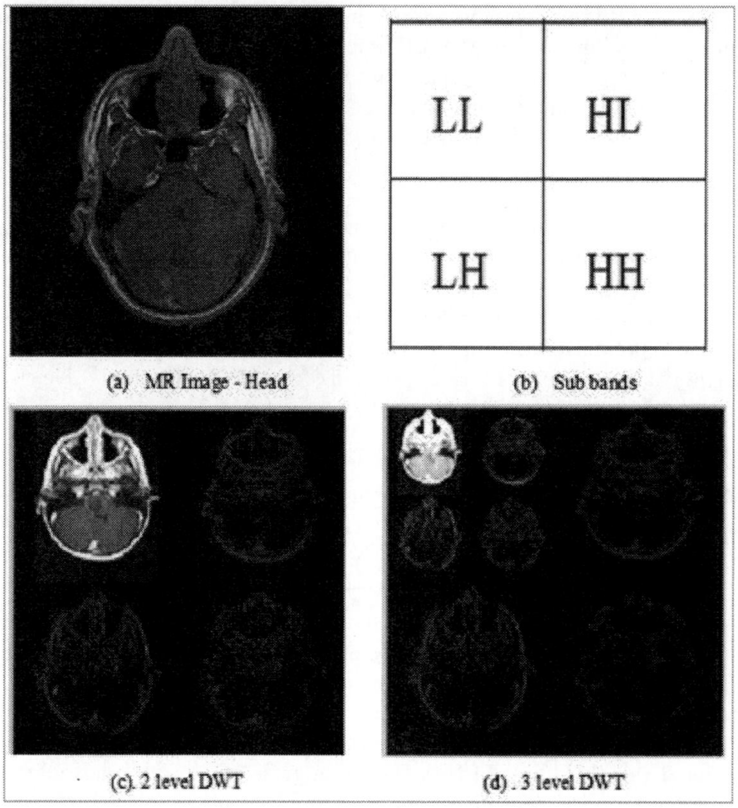


**Fig. 1.** Daubechies DWT Analysis of MR image.

### 3.3 WT Fusion Method

The multi-resolution based image fusion and the wavelet based fusion [17] are a significant method in the transform domain. The fusion rule is applied to the transformed coefficients, and the modified coefficients are used to construct the fused image in the spatial domain, using inverse transform, in all transformed domain technique.

In wavelet transform fusion, pre-registered images, m (x, y) and c (x, y), are fused to get a transform domain, the fused image F (u, v) with the help of wavelet transform $\psi$ and a fusion rule $\alpha$. The equation gives the mathematical representation of this process [18]. Similarly, the fused image in spatial domain f(x, y), is going to be reconstructed from F (u, v) with the help of inverse wavelet transform $\psi^{-1}$.

$$F(u,v) = \alpha\left(\psi(m(x,y)), \psi(c(x,y))\right)$$

$$f(x, y) = \psi^{-1}(\mathrm{F(u, v)})$$

Fig. 2 shows the block diagram representation of the process of wavelet fusion method between MR and CT images.

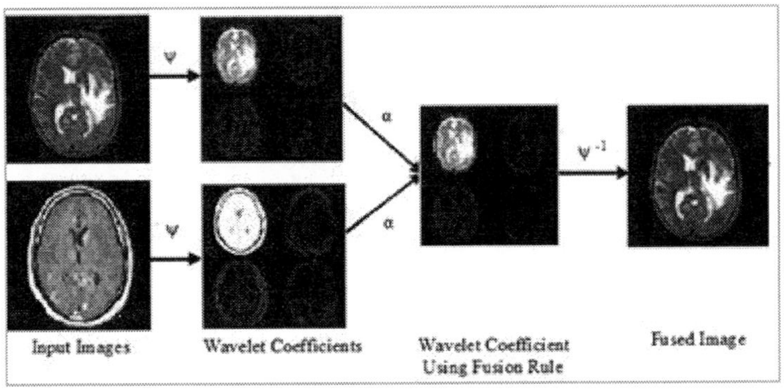


**Fig. 2.** Wavelet Transform Fusion

## 3 Proposed Image Fusion Method

Our proposed fusion method differs from the conventional wavelet method of fusion by applying KL transform prior to the wavelet decomposition. The proposed technique performs wavelet decomposition on the first principal component image instead of the whole image, which resulting good spectral resolution in the fused image. Also, the proposed scheme differs in fusion rule with conventional, by applying different fusion rule for different frequency band coefficients which resulting in good spatial resolution in the fused image. The proposed scheme of image fusion between MR and CT images is depicted in Fig. 3.

First, KL transform is performed on pre-registered input images to get the first principal component image with the reduction in dimensionality and noise. All the other principal component images belong to lower eigen values of the corresponding image are unaltered. The first principal component images from both CT and MR images which contains the most amount of information from the original images are divided into different bands of coefficients for the required levels using DWT. The fusion rule is applied separately for LL and other LH, HL and HH sub-images. Pixel averaging fusion rule was applied to the coefficients belongs to low-frequency sub-images and gradient based maximum selection scheme was applied to higher frequency band sub-images. Then the inverse wavelet transform gives the fused first PC image. Finally, inverse KL transform is performed on the fused first PC images and unaltered other PC images to obtain a higher resolution and a sharper edge image.

The complete fusion phases of our proposed method are given below.

1. KL Transform is applied to both MR and CT input images to produce the first principal component image and all the other principal component images are unaltered.
2. WT is then applied to the first PC images to get LL, HL, HH, and LH sub-band images.
3. The sub-band image coefficients are combined with a different fusion rule.
4. The synthesis wavelet function is performed to the fused sub-band coefficients from step 3 to form the fused first PC image.
5. Finally, the inverse KL transform is applied at the fused first PC images from step 5 and unaltered other PC images from step 1 to obtain a higher resolution and a sharper edge image.

The complete algorithm stated above is depicted as the flowchart in Fig. 3, and the rule is shown in Fig. 4. In Fig. 3, M (x, y) and C (x, v) represent the first principal component images of MRI and CT images respectively. hM (x, y) and h C(x, y) represent the HH band of sub-image of M (x, y) and C (x, y) respectively. IM (u, v) and IC (x, y) represent the other bands of sub-images of M (x, y) and C (x, y) respectively.

In the fusion rule, the coefficients of the LL band are fused by the average fusion rule while the coefficients of the other LH, Hl and HH bands are fused with a gradient-based maximum selection scheme.

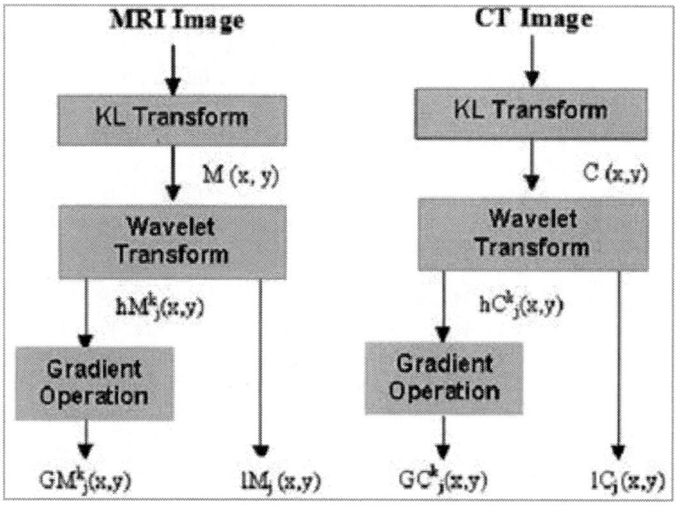


**Fig. 3.** Image Decomposition Using Wavelet Transform.

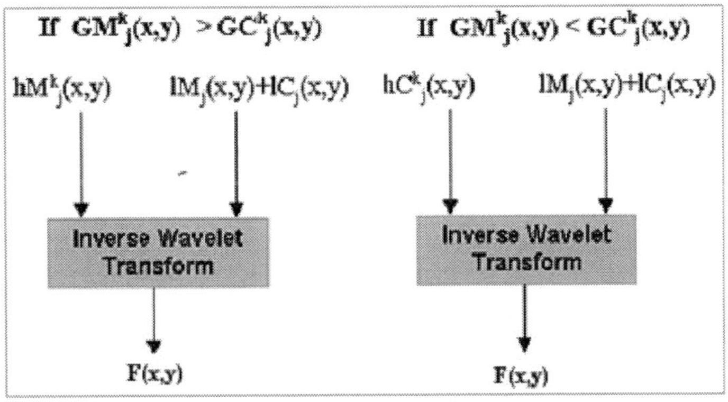


**Fig. 4.** Fusion Rule.

## 4 Evaluation Techniques

Both spatial and spectral quality measure are used for the quantitative assessment of the proposed fusion technique. The analysis indicates that our proposed algorithm, PCA-based wavelet technique for image fusion scheme has the best measure of spectral and spatial qualities.
In our assessment, the parameters like ACC, SD, and RMSE are used for the evaluation of the spectral quality of fused images.

ACC implies the correlation between each band of the fused image and reference MRI, CT images. If the value of correlation is high, it indicates lower spectral distortion [19].

The RMSE gives the difference intensity values between two images. The lower the value of RMSE indicates a better performance of fusion. RMSE is calculated for M X N images using the following formula.

$$RMSE = \frac{1}{MN}\sum_{i=0}^{M-1}\sum_{j=0}^{N-1}(R(i,j) - X(i,j)^2$$

SD is a second-order measure of the relative deviation of image contrast with respect to the first-order statistic 'mean' μ. Better the value of standard deviation, the spatial content is superior. The mathematical expression for the mean and standard deviation are given below.

$$\sigma^2 = \frac{1}{MN}\sum_{i=0}^{M-1}\sum_{j=0}^{N-1}(X(i,j) - \mu)^2$$

$$\mu = \frac{1}{MN}\sum_{i=0}^{M-1}\sum_{j=0}^{N-1} X(i,j)$$

For the purpose of assessment, the parameters PSNR, entropy, and AG are used for the evaluation of the spatial quality of fused images [20].

The amount of information in an image can be measured by entropy, and the higher value indicates more information content. If the probability distribution of each gray level is given as p= {p0, p1, p2...... pL-1} with 'L' total gray levels, then the value of entropy is calculated by

$$H = -\sum_{i=0}^{L-1}(P_i x \log_2 P_i)$$

The average gradient is a measure to access the amount of spatial resolution in an image which reflects the clarity of that image. Better the value of the average gradient, the spatial content is superior.

The AG of an image is calculated by the following equation.

$$AG = \frac{1}{(M-1)(N-1)}\sum_{m=1}^{M-1}\sum_{m=1}^{N-1}\sqrt{\left[\left(\frac{\partial f(m,n)}{\partial m}\right)^2 + \left(\frac{\partial f(m,n)}{\partial n}\right)^2\right]\Big/2}$$

PSNR is also another spatial content measure of an image based on RMSE and its maximum gray value '$f_{max}$'. The higher value indicates the better fusion performance. It is calculated using the formula:

$$PSNR = 10\ \log(\frac{255^2}{RMSE^2})$$

## 5 Experimental results and Discussions

The experiment was performed on four sets CT and MRI images of male Head, Brain, Thorax, and Pelvis for accessing the performance of our fusion technique. The algorithms are coded in MATLAB Version: 7.14.0.739 (R2012a). Daubechies wavelet is applied with three levels on CT and MRI image sets which are obtained from the U.S. National Library of Medicine. Fig. 5 shows the CT and MR image sets. Their statistical values such as entropy, average gradients, and standard deviation are given in Table 1. The performance of the proposed method is compared with three other methods, Averaging, Pyramid and DWT.

**Table 1.** The quantitative measure of the original CT and MRI image pair

| Metric | Image Set 1- Head | | Image Set 2- Brain | | Image Set 3- Thorax | | Image Set 4- Pelvis | |
|---|---|---|---|---|---|---|---|---|
| | CT | MRI | CT | MRI | CT | MRI | CT | MRI |
| Entropy | 4.512 | 3.698 | 6.633 | 2.225 | 6.115 | 4.848 | 6.470 | 3.970 |
| AG | 4.003 | 1.843 | 5.357 | 2.853 | 4.297 | 2.001 | 4.897 | 1.230 |
| SD | 37.21 | 31.12 | 57.12 | 36.56 | 30.06 | 29.22 | 36.01 | 30.09 |

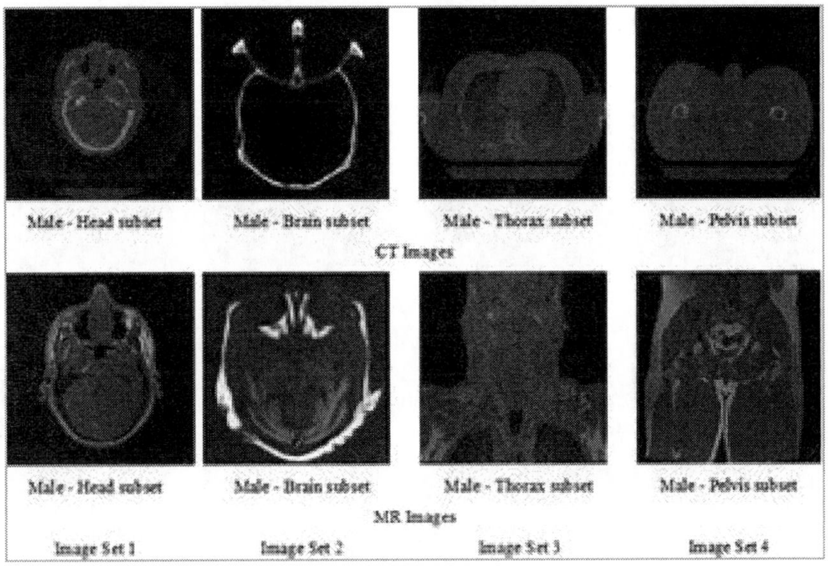


**Fig. 5.** Test Image Sets

The experimental result of fusion for the CT-MR image sets is given in Fig. 6. Each row of Fig. 6, provides the fused images of four sets of CT- MR Image pair. It is ordered as pixel averaging, the pyramid, wavelet and our proposed scheme of fusion methods. It is evident from the Fig. 6 the fused image of our scheme looks best among all methods in clarity with sharp edges.

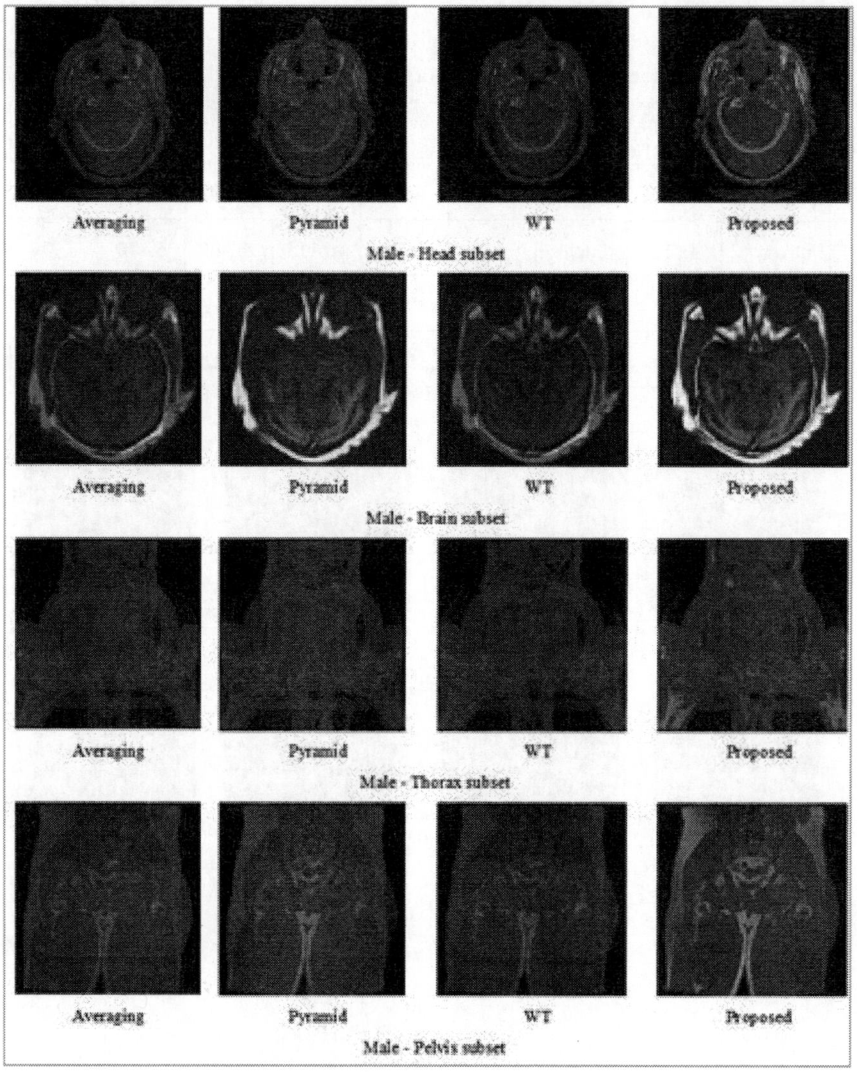


**Fig. 6.** Fused Images

Fig. 7 shows the comparison of the entropy, AG and SD of CT, MR and our fused images of four dataset. It is clearly showed that the values of these three parameters are higher in our proposed fused images than CT and MR images. So our proposed method of fusion technique comprises complementary information of both CT and MR images.

The spectral and spatial quality measures for pixel averaging, the pyramid, wavelet, and our proposed fusion technique, are indexed in the Table 2. It is evident from Table 2, Fig. 8 and Fig. 9, that the average correlation coefficients, standard deviation, the average gradient, entropy, PSNR of our fused images are higher than the other three methods of fusion. The RMSE value of our method is lower than the other methods of fusion. This lower RMSE value of our scheme is the better measure of the effectiveness of our fusion scheme.

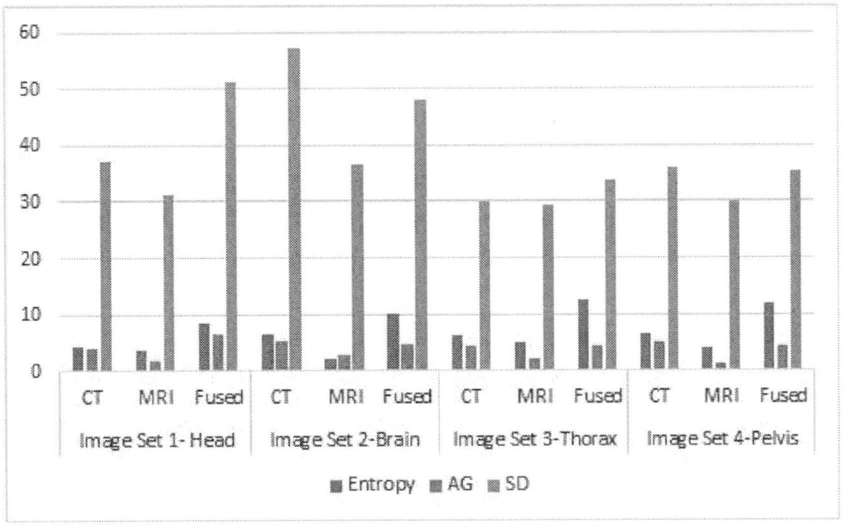


**Fig. 7.** Comparison of the entropy, AG and SD of CT, MR and our fused images

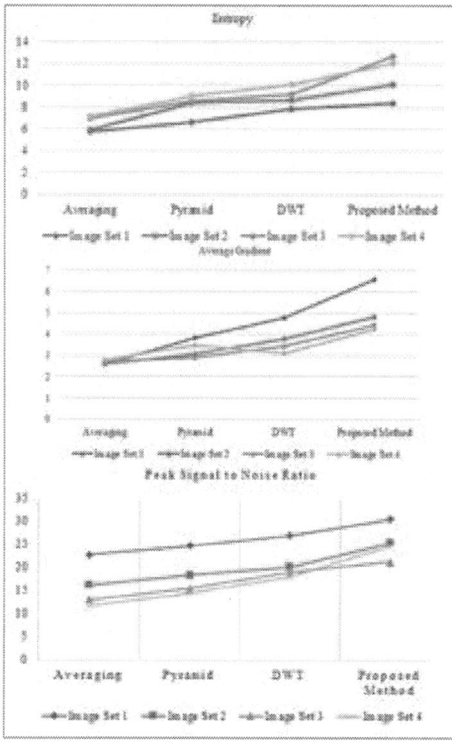


**Fig. 8** Spectral Quantity Assessment

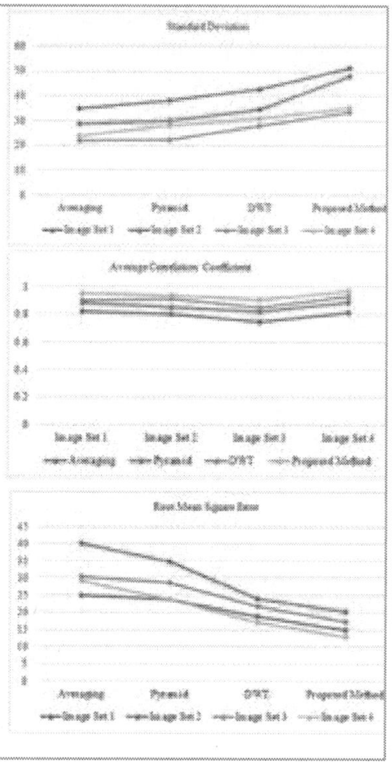


**Fig. 9** Spatial Quantity Assessment

**Table 2.** The quantitative evaluation results

| Type | Parameter | Method | Image Set 1 | Image Set 2 | Image Set 3 | Image Set 4 |
|---|---|---|---|---|---|---|
| Spectral Quality Assessment | ACC | Averaging | 0.8216 | 0.8007 | 0.7474 | 0.8152 |
| | | Pyramid | 0.8839 | 0.8545 | 0.8203 | 0.8914 |
| | | DWT | 0.9042 | 0.9112 | 0.8494 | 0.9328 |
| | | Proposed Method | 0.9481 | 0.9393 | 0.9064 | 0.9716 |
| | RMSE | Averaging | 24.9266 | 40.0056 | 30.4261 | 29.1402 |
| | | Pyramid | 23.7128 | 34.6452 | 28.6027 | 23.7557 |
| | | DWT | 18.9266 | 24.0056 | 21.7936 | 17.1402 |

| | | | | | | |
|---|---|---|---|---|---|---|
| | | Proposed Method | 15.0412 | 20.4142 | 17.3387 | 13.0157 |
| | SD | Averaging | 34.7332 | 28.7332 | 21.9033 | 23.9373 |
| | | Pyramid | 38.1163 | 30.1163 | 22.1955 | 28.1846 |
| | | DWT | 42.7284 | 34.7284 | 27.9068 | 30.9307 |
| | | Proposed Method | 51.0598 | 48.0598 | 33.6747 | 35.3494 |
| Spatial Quality Assessment | Entropy | Averaging | 5.6995 | 5.8981 | 6.9736 | 7.1262 |
| | | Pyramid | 6.5832 | 8.4271 | 8.6057 | 8.9997 |
| | | DWT | 7.8643 | 8.6800 | 9.1796 | 10.0734 |
| | | Proposed Method | 8.371 | 10.1291 | 12.6743 | 12.0508 |
| | AG | Averaging | 2.6045 | 2.5944 | 2.7172 | 2.7861 |
| | | Pyramid | 3.839 | 3.0616 | 2.9056 | 3.4795 |
| | | DWT | 4.7893 | 3.7944 | 3.4623 | 3.1199 |
| | | Proposed Method | 6.568 | 4.8013 | 4.4472 | 4.241 |
| | PSNR | Averaging | 22.6234 | 16.1224 | 12.9611 | 11.6494 |
| | | Pyramid | 24.6192 | 18.2678 | 15.3845 | 14.4079 |
| | | DWT | 26.9764 | 20.1124 | 18.9789 | 18.0238 |
| | | Proposed Method | 30.4518 | 25.2817 | 21.2309 | 24.5804 |

The fusion of MR and CT medical image gives combinational data in a single image which is fundamental to the clinical conclusion and radiotherapy. In this research, a novel PCA based wavelet method for fusing MR and CT images is proposed, which involves five phases. In the initial step, KL transform is applied to both CT and MR images to create different principal component images, according to the eign values. In the second step, first principal component images of both CT and MR images are decomposed into sub-images for the required levels by applying discrete wavelet transform. In the third step, the average fusion rule is applied to the coefficients of the low frequency band and gradient weighted maximum fusion rule is performed to the coefficients of the high frequency band. In the fourth step, the fused first PCA image is computed by inverse wavelet transform with the composite coefficients. In the last step, the reverse KL transform is performed on the fused first PC image and unaltered other PC images to acquire a higher resolution and a sharp edged image. The spatial and spectral resolution of our fused

## 6 Conclusion

Images are quantitatively assessed by two sets of metric measure. The spectral quality is assessed by the average correlation coefficient, root mean square error, and the standard deviation. The spatial quality evaluation is gotten to by the peak signal to noise ratio, entropy, and an average gradient, and Peak Signal to Noise Ratio. The investigation demonstrates that our proposed methodology, the PCA-based wavelet method of image fusion scheme has the satisfactory measure concerning spectral and spatial features. The fused image also preserved greater valuable details with significant spatial resolution.

An integrated algorithm for medical image fusion of CT and MR images is devised based on the principal component analysis in the wavelet domain. This approach enhances both spectral and spatial quality of the fused image, which would be useful for image-guided brain surgery. There is a scope to evaluate and adopt empirical mode decomposition of signal decomposition and analysis, instead of PCA, to fuse the images obtained from different sensor modalities. There is also a possibility to use empirical mode decomposition in image fusion for image set other than CT and MR modalities.


## Acknowledgment

This work was supported, in part, by Burydah Private Colleges. Special thanks to the management of Burydah Colleges.


## References


1. V. Bhateja, A. Moin, A. Srivastava, A. Lay-Ekuakille, D. N. Le.: Multispectral medical image fusion in Contourlet domain for computer based diagnosis of Alzheimers disease. Review of Scientific Instruments. Vol. 87, No: 7, (2016).
2. Hyder Ali. S., Dr.(Mrs). Sukanesh. R.: An Efficient Algorithm for Denoising MR and CT Images Using Digital Curvelet Transform, Springer Advances in Experimental Medicine and Biology – Software Tools and Algorithms for Biological Systems , Volume 696, Part 6, pp 471-480. (2011).
3. Maruturi Haribabu, CH Hima Bindu, Dr. K. Satya Prasad.: Multimodal Image Fusion of MRI-PET Using Wavelet Transform. In: IEEE International Conference on Advances in Mobile Network, Communications and its Applications. (2012).

4. Bhavana.V, Krishnappa H.K.: Multi-Modality Medical Image Fusion - A Survey. International Journal of Engineering Research & Technology.Vol:4, Issue: 02, P. 778-781, (2015).
5. Nikolaos Mitianoudis, Tania Stathaki.: Pixel-based and region-based image fusion schemes using ICA bases, Journal Information Fusion, Volume 8 Issue 2, Pages 131-142, (200.
6. S. G. Mallat.: A theory for multi-resolution signal decomposition: the wavelet representation. IEEE Transaction on Pattern Analysis and Machine Intelligence. vol. 11. No. 7. PP. 674-693, (1989).
7. Changtao He, Quanxi Liu, Hongliang Li, Haixu Wang.: Multimodal medical image fusion based on IHS and PCA. In: Procedia Engineering. Volume 7, Pages 280-285, (2010).
8. Yong Yang, Dong Sun Park, Shuying Huang, Zhijun Fang, Zhengyou Wang.: Wavelet based Approach for fusing Computed Tomography and Magnetic Resonance Images. IEEE *Chinese Control and Decision Conference*. pp. 5770-5774, (2009).
9. Fatma El-Zahraa Ahmed El-Gamal, Mohammed Elmogy, Ahmed Atwan.: Current trends in medical image registration and fusion, Egyptian Informatics Journal, Volume 17, Issue 1, Pages 99-124, (2016).
10. Padmavathi K, Mahima Bhat, Maya V. Karki. : Feature Selection based on PCA and PSO for Multimodal Medical Image Fusion using DTCWT. CoRR, (2017).
11. Aisha Moin, Vikrant Bhateja, Anuja Srivastava. : Weighted-PCA Based Multimodal Medical Image Fusion in Contourlet Domain. In: Proceedings of the International Congress on Information and Communication Technology. Advances in Intelligent Systems and Computing, vol 439, Springer, Singapore, (2016).
12. Yong Yang, Dong Sun Park, and Shuying Huang, Nini Rao.: Medical Image Fusion via an Effective Wavelet Based Approach, EURASIP Journal on Advances in signal processing, Volume 2010, Article No: 44, (2010).
13. Chithra Chandran, C. Jyothirmayi Dev.: A Modified Three Level Fusion Technique for Multimodal Medical Image Fusion in Wavelet and NSCT Domain. International Journal of Scientific & Engineering Research, Volume 6, Issue 5, pp 259-263, (2015).
14. Yong Yang.: Performing Wavelet Based Image Fusion through Different Integration Schemes. In: International Journal of Digital Content Technology and its Applications. Volume 5. Number 3. (2011).
15. Yosef Jbara, Hyder Ali. S.: An efficient Artifact Free Denoising Technique for MR Images relying on Total Variation based Thresholding in Wavelet Domain. Graphics, Vision and Image Processing Journal, Volume 18, Issue 1, ICGST LLC, Delaware, USA, (2018).
16. Yong Xia, Zexuan Ji, Andrey Krylov, Hang Chang, Weidong Cai.: Machine Learning in Multimodal Medical Imaging. BioMed Research International, vol. 2017, Article ID 1278329, 2 pages, (2017).
17. L.J. Chipman, T.M. Orr, L.N. Lewis.: Wavelets and image fusion. *IEEE Transactions on Image Processing*, 3, pp 248–251, (1995).
18. Z. Omar, S. S. Ahmed, M. Mokji, M. Hanafi, V. Bhateja.: Wavelet-based medical image fusion via a non-linear operator. IEEE Region 10 Conference (TENCON), Singapore, pp. 1262-1265, (2016).
19. Li, S., Yang, B., Hu, J.: Performance Comparison of Different Multi-resolution Transforms for Image Fusion. Information Fusion. 12(2). pp 74–84. (2011).
20. Bindu, C. H., Prasad, K. S.: Performance Analysis of Multi Source Fused Medical Images Using Multi-resolution Transforms. In: International Journal of Advanced Computer Science and Applications. Vol 3. pp 54–62. (2012).

## Appendix

### Nomenclature

| | |
|---|---|
| PCA | Principal Component Analysis. |
| KL | Karhunen-Love. |
| CT | Computed Tomography. |
| MR | Magnetic Resonance. |
| ACC | Average Correlation Coefficient. |
| PSNR | Peak Signal to Noise Ratio. |
| RMSE | Root Mean Square Error. |
| AG | Average Gradient. |
| SD | Standard Deviation. |
| MRA | Magnetic Resonance Angiogram. |
| PET | Positron Emission Tomography. |
| SPECT | Single Positron Emission Computed Tomography. |
| HVS | Human Visual System. |
| DWT | Discrete Wavelet Transform. |
| LL | Low – Low. |
| LH | Low – High. |
| HL | High – Low. |
| HH | High – High. |

### Biographies

Dr. Yosef Hasan Jbara is an Assistant Professor in the College of Engineering and Information Technology at Buraydha Colleges. He received his Ph.D. in Artificial Intelligence from the Faculty of Computer Information Systems, University of Banking and Financial Sciences. He has published in the areas of artificial intelligence; simulation modelling; image processing and data mining. He has published several research papers in International Journals and conferences. His email addresses are yosef.hasan@bpc.edu.sa and yosefjbara@ieee.org.

# Mining Knowledge for Differential Diagnosis of Pulmonary Nodules

P. Perner[1], T.P. Belikova[2], I. Stenina[2]

[1]Institute of Computer Vision and Applied Computer Sciences IBaI, Germany,
[2]Institute for Information Transmission Problems Russian Academy of Sciences, Moscow, Russia
pperner@ibai-institut.de

**Abstract.** Development of computer-assisted radiology in PACS is associated with the use of new intelligent capabilities of Computer-Assisted Radiology. In this paper we present our work on data mining to support medical image diagnosis. We use decision tree induction in order to learn the expert knowledge, presented in the form of image descriptions in the database. A methodology is developed to perform data mining in picture archiving systems. The tool for data mining is presented. It was applied in the task of early differential diagnosis of pulmonary nodules in lung tomograms and allowed to support image analysis and classification by the expert. Results of knowledge acquisition on the basis of data mining are presented and compared with the results of the semi-automatic method of knowledge acquisition by a Syndrome framework tool. The developed decision tree and developed Syndrome-like decision rule were effective for early diagnosis of peripheral lung cancer, so that we are able to apply this method to other medical tasks, in particular, for image analysis in mammography.

**Keywords:** Data Mining, Decision Support, Image Interpretation, Decision Trees, Syndrome Classifier, Differential Diagnosis, Pulmonary Nodules

## 1 Introduction

Nowadays radiology departments are in the center of radical changes in technologies of acquisition, storage, transmission, and analysis of diagnostic images. The radiographic films, which have been used for image analysis since 1895, are being replaced now by digital images acquired by new imaging modalities (CT, MRI, etc. [1][2]). Picture Archiving and Communication Systems (PACS) were developed to provide efficient and cost-effective analysis, exchanging, archiving, and retrieving of diagnostic images presented in digital form [2]-[4]. It opens new extended possibilities for manipulation, processing, and handling medical data by computers to improve the efficacy of medical investigations. PACS gives the user means to surpass current diagnostic capability thanks to the achievements of Computer-Assisted Radiology such as multi-modality imaging and multimedia displaying of medical data, image processing and computer-assisted decision-making [5]-[7].

Further development of computer-assisted radiology is associated with the use of new intelligent capabilities, such as data mining and automatic image analysis, which are being developed to support decision- making by the expert.

Medical image diagnosis involves several task solutions: a) lesion detection; b) recognition of diagnostically important features (structures, details, objects, etc.); so also c) interpretation of the image as a whole. Solution of all these tasks requires a long practice and specific domain knowledge for image features analysis and scene understanding. In the most interesting and practical cases, such as early stage diagnosis, the physician has to detect tiny changes in the complex image background, to analyze the lesions, which have no obvious specific features, and to make differential diagnosis on the basis of noisy, incomplete, and unbiased data, obtained as a result of image analysis.

Image processing could promote image analysis by more explicit imaging of specific features. The application of data mining can help to obtain knowledge about specific features of different classes, and to create models for decision-making. It can also provide the discovery of some inherent non-evident links between classes and their imaging in the picture.

The goal of this work was the development of knowledge acquisition methods for medical image diagnosis, based on data mining, which could help to solve some cognitive, theoretical and practical problems:

1. to find features, which are basic for human reasoning and classification and to discover the tasks of expert's purpose vision.
2. to reproduce and to display a decision model for specific medical task solution, which can be used as a tool to support decision-making by physicians, who are not experts in the specific field of knowledge. Such a model could be also useful for teaching novices.

The second goal of the work was the development of methodology for the application of data mining in medical picture archiving systems. We applied knowledge acquisition methods to the task of early differential diagnosis of solitary pulmonary nodules in lung tomograms. In Section 2, we describe the recent approaches to knowledge acquisition in different fields of knowledge. Materials and the developed method are described in Section 3. Results are given in Section 4. We compare facilities of the developed automatic and semi-automatic knowledge acquisition methods and give the results in Section 5. Finally, we summarize our experience in Section 6.

## 2 Knowledge Acquisition Methods

Knowledge acquisition is the first step when developing a system to support image analysis and interpretation. The kind of method used for knowledge acquisition depends on the inference method, which the system is based on [9]-[14].

The knowledge acquisition process for a rule-based system is usually manually done by interviewing a human expert [9] or by employing interactive knowledge acquisition tools, e.g., repertory grids [10].

In model-based systems, the knowledge representation is based on semantic nets. These structure the domain knowledge into concepts and their relations. The language of semantic nets determines the way in which new knowledge is elicited.

When generalized knowledge is lacking, then case-based reasoning [13] seems to be a proper method. An interpretation of such a system is based on determining the case (or cases), stored in the case base, which are the closest to the actual case. The measured value of closeness and the interpretation, associated with the similar case in the case base, is presented to the user. How to interpret the closeness is left to the user. A case-based reasoning system developed for image interpretation is described in [14]. The limited explanation capability is the main lack of case-based reasoning systems-

We are developing a knowledge acquisition method for those applications, where no generalized domain knowledge is available, but there is a large database of images, associated with expert description and interpretation. This task is very up to date taking into account the recent trend in the wide use of picture archiving systems in Medicine and in other domains.

The application of data mining could help get some additional knowledge about specific features of different classes and variations of their manifestation on the image.

Such a system opens possibilities for automated expert-independent image analysis based on measuring some specific features directly in the image.

## 3 Materials and Methods

For our experiment, we used a database of tomograms with verified diagnoses. Annual screening by means of radiography identified all studied patients with pulmonary nodules. Patients with pulmonary nodules up to 3 cm in size were selected (Set 1: 218 cases =80 benign + 138 malignant (peripheral lung cancer)). Patients with clinically verified cancer in other organs were excluded from the test to avoid cases of metastases in lung.

For our test experiment we selected 38 images (Set 2: 38 cases=20 malignant+18 benign). About a half of these images were referred to as complex cases as they yielded ambiguous diagnostic decisions during the analysis of the unprocessed images by the expert.

### 3.1 Acquisition of Lung Images

Conventional (linear) coronal plane tomograms with 1-2 mm section thickness were used for specific diagnosis. The slice was centered through the nodule. One or two contiguous sections with 3 mm advance between the slices have been done for some cases as well.

Original linear tomograms were digitized with step of 100 micron (5.0 line-pairs per millimeter) to get 1024 x 1024 x 8 bits matrices with 256 levels of gray. Such a digitization promoted an acquisition of images with high spatial resolution of anatomical details that were necessary for the specific diagnosis of lung nodules.

### 3.2 Enhancing Diagnostically Important Details

A method of digital filtering [15] was applied to improve the specific diagnosis of small solitary pulmonary nodules in lung tomograms. The goal of processing was to emphasize the specific nodule features and to make them more distinct. The fulfilled steps involved:

1. Modeling the mixture of the background and the useful signal for the observing image set [16][17];
2. Development of optimal filter to suppress the impact of the background and to emphasize the remaining part of the image, which contained informative details [17];
3. Modeling the background; which fits to the considered class of images [17];
4. Testing the processed images by the experts to select the background model, which allows the best imaging of diagnostic structures and the highest efficacy of image interpretation by the physician. The feedback was used to find the optimal filter, which met the expert criteria, and gave the best feature presentation for expert's purpose vision [17][18].

X-ray - morphological comparison was fulfilled to confirm that all details, displayed in the processed tomogram, corresponded to real morphological structures found by morphologists in post-surgical histotopograms [17][18].

The processing improved manifestation of diagnostically important details on medical images and thus helped the physician to be more certain in feature reading and interpretation.

However, in diagnostically complex cases good feature imaging was insufficient for correct image interpretation by the radiologist. It was caused by the absence of evident specific features at the early stage of disease, and by insufficient knowledge for making decisions on the basis of incomplete and fussy data, obtained during image analysis. We applied data mining methods to support decision-making by the physician on the basis of the observed complex of features.

### 3.3 Acquisition of Expert Concepts for Image Description

First, together with the expert we developed an attribute list, which the radiologist applied for image description. The attribute list covered all concepts about diagnostically important features and feature values, which the expert used to use during image analysis and interpretation.

The expert described each processed image from our set of images in the terms of selected attributes, which were then collected in symbolic form in the database of image descriptions.

We tested two variants of Attribute lists, which differed in the presentation of feature values [19] (see Attribute List (1), Table1 and Attribute List (2), Table 2).

### 3.4 System for Image Processing and Analysis

Experiments with digital image processing and analysis were carried out using the workstation developed in the Institute for Information Transmission Problems Russian Academy of Sciences [17]. The Workstation comprised of a PC computer with hard and floppy discs, a device for image input (laser drum scanner or plane-table scanner), a high-resolution color monitor and printer.

The radiologist worked as an expert in this system. He communicated with a computer answering to its requests: determined and outlined the area of interest and the area of the nodule. The parameters of the optimal filter were then calculated automatically [17]. The radiologist watched the processed image (Fig. 1, Fig. 2), displayed on-line on a TV monitor, evaluated its specific features (character of boundary and shape of the nodule, specific objects, details and structures inside and outside the nodule, etc.) according to the list of attributes (Table 1 or Table 2) and inputted the codes of appropriate attribute values into the database program. Uncertain and missed values were available in the database as well. Hard copies of the previously processed images from the archive have also been used in this work.

Two radiologists of different qualification levels took part in our experiments. The first one was the 'Expert' of high skill level, who had a long experience in image reading and interpretation. The second one was a pulmonologist, who was not trained to read processed images. So, he was a 'Novice' in the interpretation of details in post processed tomograms. Each radiologist analyzed the processed images and inputted appropriate attribute values into the database or missed it if he was not sure what value fitted to the case.

The collected data set was then passed as a dBase-file to the inductive machine-learning tool [20] and to Partner System [21] for Syndrome-like decision rule creation.

**Fig. 1.** Workstation for image processing and analysis

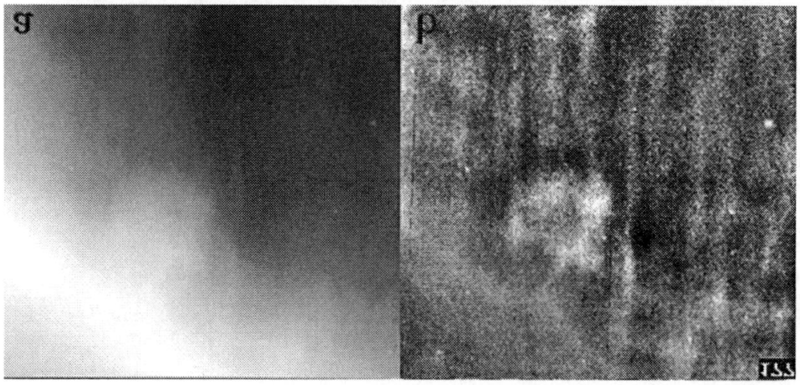

**Fig. 2.** Malignant nodule a) A fragment of the original tomogram with a shadow of nodule. b) Processed tomograms with enhanced imaging of nodule features. After processing one can see non-round nodule with, non-smooth irregular margin. There are irregular areas of varied densities inside the nodule. Multiple shadows of vessels are at the nodule, converging to it.

### 3.5 System for Data Mining

For the data-mining experiment we used the tool ***DECISION_MASTER***, developed in the Institute for Computer Vision and applied Computer Sciences IBaI [20] (Leipzig, Germany). The tool is written in C++ and operates in Windows95 or Windows

NT. It has a user-friendly interface and can be easily used by non-specialists in computer science.

The developed tool can create binary and n-ary decision trees from the data, presented in the dBase-file. It has several options allowing one to specify how numerical features should be partitioned [22] and what method should be used for feature selection. Evaluation of the results can be done by test-and-train and n-fold cross-validation [23]. Missed values can be handled by different strategies [24]. The tool also provides functions for outlier detections.

For our experiment, we used a decision tree induction method, which created binary-trees based on maximum entropy criteria [25]. Missing value were handled by replacing them with the mean value of the corresponding class [24]. Pruning was done based on the reduced-error pruning technique [26]. We applied 10-fold cross-validation for evaluation. Besides the error rate we calculated Sensitivity for Class_1 $c_1$ and Specificity for Class_2 $c_2$, which are error criteria usual for medical applications:

$$E_{sens} = S_{c1m} / N_{c1} \qquad E_{Spec} = S_{C2m} / N_{C2} \tag{1}$$

where $S_{c1m}$ is a number of correctly classified samples in Class 1 and $N_{C1}$ is a number of all samples in Class 1 and $S_{C2m}$ and $N_{C2}$ in Class 2 respectively.

When the decision model (diagnosis knowledge) has been learnt, the rules are provided either in TXT format for further use in the expert system, or the expert can use the learnt knowledge in the diagnosis unit of Decision Master for interactive work.

The architecture of Picture Archiving System combined with the data-mining tool is shown in Fig. 3 [27]. The screenshot of our data mining tool Decision Master is shown in Fig. 4.

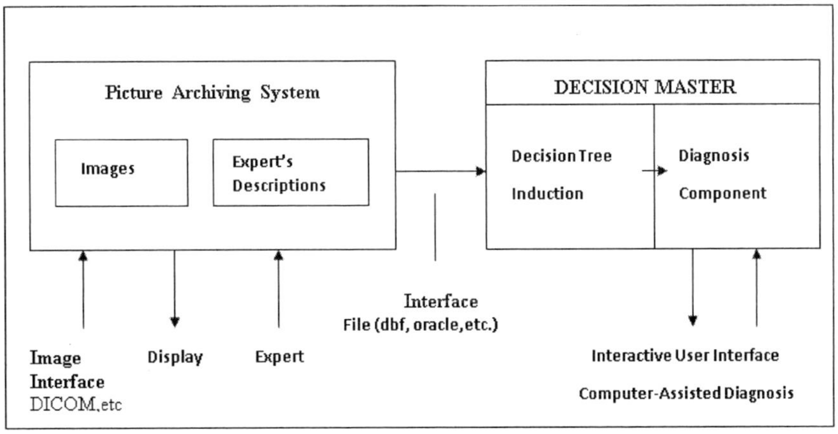


**Fig. 3.** Architecture of Picture Archiving System combined with the Data Mining Tool.

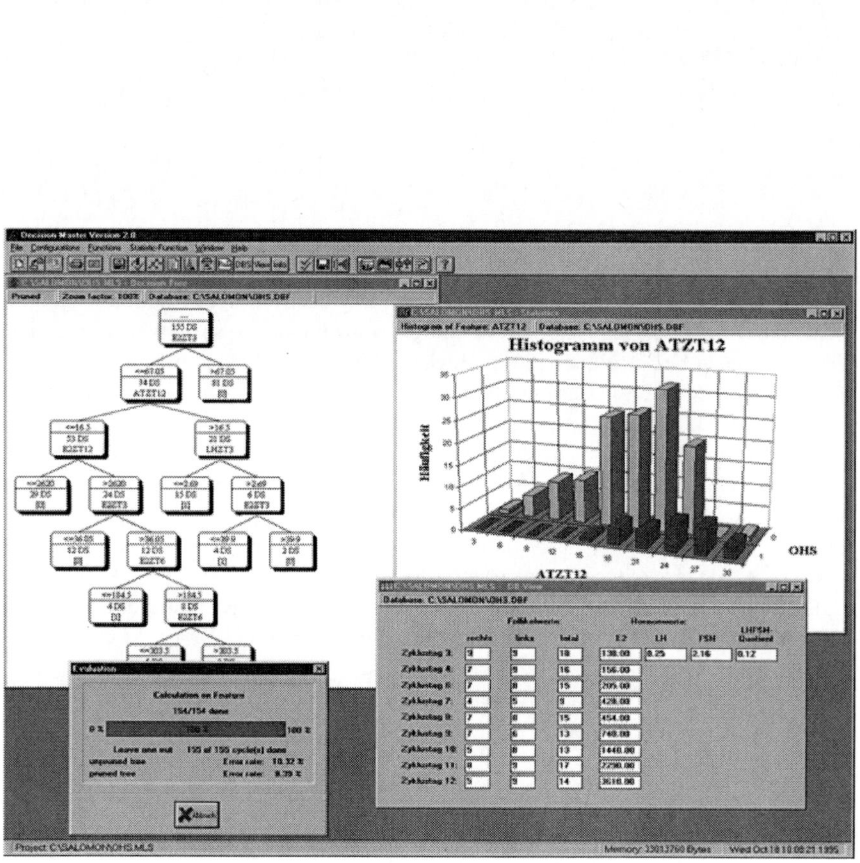


**Fig. 4.** Screenshot of Data Mining Tool ***Decision Master***

### 3.6 Semiautomatic Knowledge Acquisition for Syndrome-Framework Tools

We compared the performance of the induced tree classifier to the performance of the Syndrome classifier developed with the use of the Syndrome–Framework tool to learn diagnosis knowledge [21]. On Fig. 1 (left monitor) there is a Syndrome–Framework tool combined with the database and the system for image processing. This tool allows the user to develop a decision model, which belongs to a specific type of classifier: it is similar to Syndrome (decision rule), conventional for Medicine (see Fig. 5). Syndrome is a variant of a neuron network, which operates using a multilevel threshold voting scheme.

A Syndrome-Framework tool supports the use of multivariate contingency table analysis in combination with the analysis of classification results to create the effective decision model. It can be created automatically or semi-automatically (under the control of expert knowledge). In this case knowledge-guided data analysis is organized as an interactive process with the participation of a physician. The expert controls the results of automated feature analysis: He/she checks in what features and what feature values have been selected automatically as the basic features. He/she can correct this list according to his/her knowledge so that the new list (and discriminative values) "make sense" from his/her point of view. The obtained list has some redundancy in knowledge for improving the reliability of the decision-making model. The expert observes classification results using an analog of ROC curves displayed on a TV monitor for both classes [28]. The expert can change the threshold value according to his/her specific purpose and priorities, for example, he/she selects the threshold to get a higher number of correct decisions for cancer in the score of "false alarms" for benign nodules, as it is less dangerous for the patient than missed cancer cases. (see Fig.6).

Decision rule (Syndrome) is a set of feature conjunctions, based on a set of ***m*** binary features, selected from the initial list of features. The rule (Syndrome) is fulfilled for the object if any set of ***k*** features (from ***m*** basic features) are observed (have value 1) for the object, where $k \in [n, m]$, and ***n*** is a determined threshold. Tuning Syndrome parameters *[n, m]* is fulfilled automatically or under the control of expert knowledge to obtain the most redundant and efficient decision rule.

Constructed network allows classification of objects on the base of a fuzzy set of features, observed in the image. Being based on the voting principle, such a network is tolerant, to a large extent, to distortions, incompleteness, and uncertainty of input data. It doesn't need prior information about feature significance (or can use this information if it is available).

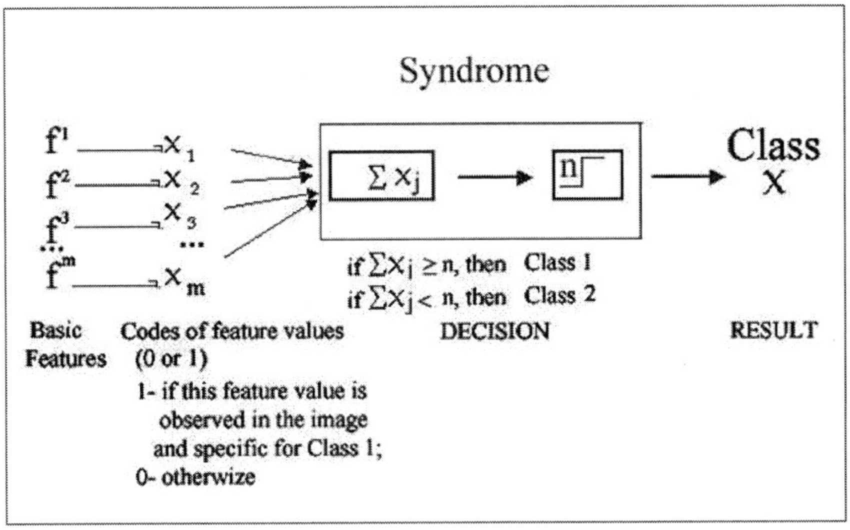


**Fig. 5.** Scheme of Syndrome decision model

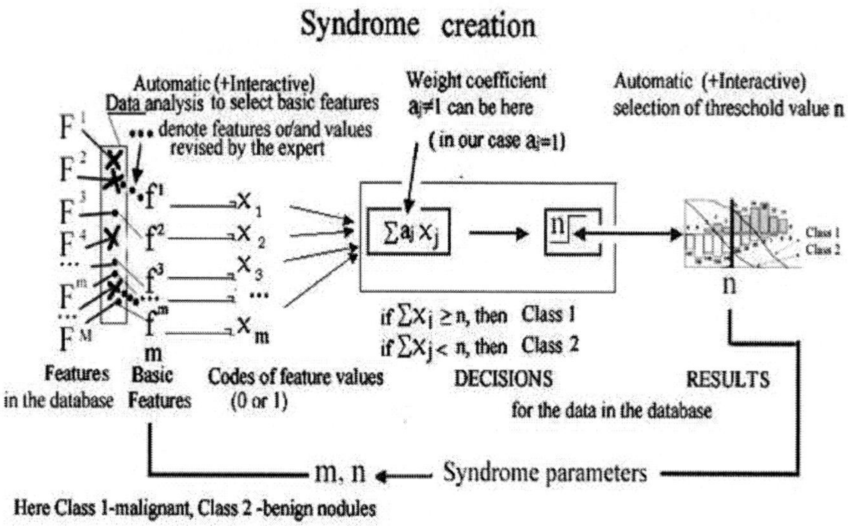


**Fig. 6.** Scheme of Syndrome creation

# 4 Results

## 4.1 Efficacy of classification for Attribute List (1)

The induced tree for Attribute list (1) is shown in Fig. 7. The unpruned tree consists of 40 leaves. The pruned tree consists of 11 leaves (see Fig. 8). The 'Expert' preferred the unpruned tree much more, because almost all attributes he used to use for decision-making appeared in this tree.

The 'Expert' confirmed that the attribute *Structure,* selected by the decision tree, is important for image analysis, as well as the attribute *Scar-like changes inside the nodule*. However, he wondered why other important features, such as *Shape* and some others didn't work for classification. The 'Expert' told us that he preferred to start nodule analysis from its *Structure* and *Scar-like changes inside the nodule*, then *Shape* and *Margin*, and, finally, *Convergence of Vessels* and *Outgoing Shadow in Surrounding tissues*. According to his opinion, these features are also very important for his final decision about the nodule nature. So, although our decision trees represented the decision in the form convenient for the final user, it did not follow the strategy of the expert.

Therefore, we looked for the error rate and used it as the main criterion of DT fitness (Table 3).

```
Scar-likeChan = 1
I  ConvofVes = 1 Class_1
I  ConvofVes = 2 Class_2
I  ConvofVes = 3
I     Margin = 1 Class_1
I     Margin = 2 Class_2
I     Margin = 3 Class_1
I     ...
I     Margin = 15 Class_1
Scar-likeChan = 2
I  Structure = 1 Class_1
I  Structure = 3 Class_2
I  ...
I  Structure = 6 Class_2
I  Structure = 7 Class_1
I  I  OutgoShadinTis = 1 Class_2
I  I  ...
I  I  OutgoShadinTis = 3 Class_2
I  I  OutgoShadinTis = 4 Class_1
Scar-likeChan = 3
I  Structure = 1 Class_1
I  Structure = 2 Class_1
I  Structure = 3 Class_2
I  ...
I  Structure = 6 Class_2
```

**Fig. 7.** Decision Tree (unpruned)

```
Scar-likeChan = 1 Class_1
Scar-likeChan = 2 Class_2
Scar-likeChan = 3
I  Structure = 1 Class_1
I  Structure = 2 Class_1
I  Structure = 3 Class_2
I  ...
I  Structure = 6 Class_2
I  Structure = 7 Class_1
```

**Fig. 8.** Decision Tree (pruned)

As it is evident from Table 4, we did not achieve the expert's diagnostic results. The reason might be in the choice of attribute values. During the tree building process the attribute, appearing first in the database and satisfying the splitting criteria, is always chosen. For some categorical attributes, there were too many categorical values in Attribute List (1). That is why during the tree building process the training set was split up into very many subsets, which had a few data samples. The tree building process stopped very soon, since remaining data set contained no samples, which could be partitioned anymore.

**Table 3.** Efficacy of classification for Attribute List (1) on data Set 1 and test Set2. Error Rate on data set and estimated error for cross-validation on sets.
(1) Evaluation on the data from Set 1 (250 cases)
(2) Evaluation of Decision tree on Test Data Set 2 (38 cases)

| | Before pruning | After pruning | |
|---|---|---|---|
| **Experiment** | **Error** | **Error** | **Estimated error** |
| **(1)** | 1,0 % | 1,0 % | 11,2 % |
| **(2)** | 11,4 % | 14,4 % | 11,2 % |

**Table 4.** Comparison of the Expert and Decision Tree (DT) classification for attribute list 1 for Set 2.

| Accuracy | | ***Sensitivity(Accuracy) for* Class_1** | | ***Specificity(Accuracy) for* Class_2** | |
|---|---|---|---|---|---|
| **Human** | **DT** | **Human** | **DT** | **Human** | **DT** |
| 94,7 % | 88,6 % | 97,2 % | 88,8 % | 91 % | 88 % |

Since the attributes are nominal, we cannot find an ordering on attribute values, therefore, we cannot summarize the values to obtain one more general attribute value.

For example, if one has attribute *intensity* with attribute values "*black, dark gray, gray, light gray, white*", it is possible to generalize values *dark gray, gray and light gray* into gray.

Shapiro proposed the approach [29] to reduce a totally opaque, large decision tree to a hierarchy of several small decision trees, each of which "made sense" to the expert. We have chosen the way of constructing a new attribute list, which allowed us to realize Shapiro's idea.

First, we can use the generalized attribute value for the tree building process. If we find in the induced tree that further distinction between the attribute values is necessary, we can carry out another induction experiment based on the specialized attribute values, starting with the data set, which corresponds to the leaf of the tree with generalized attribute value.

### 4.2 Efficacy of classification for Attribute List (2)

The initial attributes, which had numerous attribute values were replaced by a set of attributes (sub-attributes), where each sub-attribute had a lower number of attribute values -Attribute List (2), (see Table 2).

In order to makes sure that we did not develop many redundant and highly correlated attributes, we checked the reliability of features by calculating a proximity matrix based on Kruskal's tau [30] using a new data base created with the use of a new attribute list (Tab. 2). We grouped the set of features into functional groups based on an average link hierarchical clustering method [31]. We have found high correlation only between the attributes Charlung and Withlupl. For all other attributes, we were satisfied with the result. The resulting decision tree for data, collected according to Attribute List (2) is presented in Fig.8. This tree performed better, than the first decision tree (compare error rates in Table 3 and Table 5).

However, from the expert's point of view, this decision tree was difficult to interpret, as there were not enough attributes there.

**Table 5.** Efficacy of classification for Attribute List (2) on data Set 1 and test Set2
Error Rate on the data set and estimated error for cross-validation on the two sets
(1) Evaluation on data set 1 (250 cases)
(2) Evaluation of decision tree on test data set 2 (38 cases)

| **Result** | Before Pruning | After Pruning | |
|---|---|---|---|
| | **Error** | **Error** | **Estimated error** |
| (1) | 1,3% | 1,3% | 8,0% |
| (2) | 4,2% | 4,2% | 8,0% |

### 4.3 Classification results based on the test data

Therefore, it was interesting to see how the tree performed on the data of Set 1 and Test Set 2. In Table 6 there are results of DT classifier on the data, obtained by the 'Expert' and by the 'Novice'. Our experiments showed that the performance of the decision tree based on the Attribute List 2 was better in comparison with those of the high-level expert (see Table 6). The resulting error rate showed that the classifier gave a reliable error rate even for very noisy and incomplete data presented by the 'Novice' (see experiment (2) with the 'Novice' in Table 6).

**Table 6.** Comparison between Expert's decisions and classification of Decision Tree (DT): (1) High-level Expert; (2) Middle-level Expert (Novice).

| Users | *Accuracy* | | *Sensitivity* | | *Specificity* | |
|---|---|---|---|---|---|---|
| | **Class_1** | **Class_2** | **Class_1** | | **Class_2** | |
| | **Human** | **DT** | **Human** | **DT** | **Human** | **DT** |
| (1) | 94,5% | 95,8% | 96,2% | 93,75% | 90% | 100% |
| (2) | 55,2% | 73% | 61,1% | 75% | 50% | 72% |

### 4.4 Classification Results based on Syndrome Classifier

A decision rule was constructed on the basis of data obtained according to the Attribute List (2).

The initial list of features contained *N=25* features proposed by the expert. In Table 7 there is a list of specific features selected from the initial Attribute list as a result of statistical data analysis in the database and added by the expert, who selected some additional features, which were diagnostically important from his point of view [32]. These features were then used for decision rule creation. Two syndromes with different numbers m of selected basic features and threshold values n were constructed and tested [28][32]: Syndrome_1 (***m1=12, n1=4***), and Syndrome_1 (***m2=8, n2=3***). Multivariate contingency table analysis [33]was applied for this purpose. The Syndrome referred the sample to the Class_1 if any set of k from ***m*** features were observed (had value **1**) for this object, where $n \leq k \leq m$. (e.g. we must observe in the images at least ***n*** features, which have values, specific for Class1). The total number of such subrules in the rule is equal to the sum of all possible combinations $C_{km}$ of ***k*** features from m for all ***k***, which satisfies the indicated condition $n \leq k \leq m$.

A set of basic features ***m***, and a threshold value ***n*** (***m*** and *n*-Syndrome parameters) were determined so that the resulting decision model gave the highest efficacy and stability on the data in the database. Both Syndromes showed the same accuracy on data from Set_1. Both constructed Syndromes were applied for classification of Test-Set_2. Our experiments showed Syndrome_1 worked better with noisy and incomplete data from test Set (2). Therefore, it was selected for our final experiments and comparisons with decision tree results (see Tab. 8).

The developed decision rule presented effective strategies for image analysis and interpretation, orienting the user to look for specific features, and also showing how to classify the image on the basis of observed features. The rule can be easily presented in the form of Syndrome in the language that is natural for the expert's domain knowledge. Such a rule can be easily learned by the physician, applied in clinical practice or used for teaching novices.

**Table 8.** Comparison results of Syndrome-classifier and Decision Tree Classification on the database created with use of Attribute List (2)
(1) Data Readings from high-level Expert (2) Data Readings of Non-trained Novice

| *Accuracy* | | | *Sensitivity* | | | *Specificity* | | |
|---|---|---|---|---|---|---|---|---|
| Radiologist's Diagnosis | Syndrome Classifier | • Decision Tree | **Class_1** | | | **Class_2** | | |
| **RD** | **SC** | **DT** | RD | SC | **DT** | RD | SC | **DT** |
| (1) 94,5% | 95,6% | 95,8% | 97.2% | 98,4% | 93,7% | 91% | 91,7% | 100% |
| (2) 55,2% | 84,2% | 73,0% | 61,1% | 78,0% | 75,0% | 50% | 90% | 72% |

# 5 Comparison of Data Mining Method with Semiautomatic Knowledge Acquisition Method by SYNDROME–FRAMEWORK Tool and Automatic Method based on Decision Tree Induction

The test based on image readings from a high-level expert shows that the decision tree method performs better than the Syndrome (Table 8). One can see that both classifiers exceed the results, obtained by the 'Expert'. Only in the case of the 'Novice', whose data contained many missed and distorted (noisy) values, did the Syndrome perform better. Our investigation showed that in the decision tree classifier a misclassification of samples is mostly based on incorrectly chosen attributes, not on the missed attribute value, since the decision tree classifier is based only on a few attributes that are the most important and always appear in the image.

Syndrome1 uses 12 features and Syndrome2 uses 8 features. If we compare the set of features used by the decision tree with those used by the syndrome we can see a large overlap among the two feature sets. However, the decision tree induction method tries to find a minimal set of features necessary for a decision. The feature 'redundancy' in the Syndrome classifier might be a reason that it works so well for uncertain cases. Beyond that the decision tree was learnt based on a data set created by a top-level expert. Therefore, the decision tree cannot have a strategy to deal with the reasoning behavior of a middle-level expert.

In decision tree induction we cannot easily update the tree without rebuilding the entire tree. However, there is some research going on for incremental decision tree induction [33], but that is not the subject of this paper.

It is interesting to note, that decision-tree and the Syndrome–like decision rule allow one to find basic diagnostically important features. They both can support decision-making on the basis of features reading. It means that they give the user decision models, which help to decompose a complex pattern recognition task, which usually needs specific domain knowledge and long training, into a set of simpler subtasks connected with analysis of object features (in our case these were: shape, margins, homogeneous- inhomogeneous area, etc.). Most of these tasks are under consideration in the frame of Computer Vision approach. Feature computing can help to measure necessary feature values directly in the image thus producing expert-independent results of image analysis. Our next steps will be to automate the estimation of object shape and margins.

## 6 Conclusion and Further Work

In this paper we presented our methodology for data mining in picture archiving systems. The basis for our study is a large database with images and expert descriptions. Such databases result from the common use of picture archiving systems in medicine.

Applying the data mining method to the database of image descriptions we could learn the important attributes necessary for image analysis. Acquiring basic features and developing effective decision models, such methods could help to achieve the goals of purpose vision. It could help decompose a complex visual task into several simpler subtasks, present effective strategies for image analysis and interpretation, show the way for development of unbiased expert-independent methods of measuring some basic features directly in the image. They can support image analysis and interpretation by the expert in the case of uncertainty.

We showed how the domain vocabulary should be set up in order to get good results and which techniques could be used in order to check reliability of the chosen features.

Experiments have demonstrated the attributes, included automatically into the tree, represented by the expert knowledge.

We compared the decision tree and Syndrome-like classifier. The error rate for decision tree classifier was better than the error rate of the Syndrome for the data obtained from the expert. Only for handling very uncertain data was the performance of the decision tree worse than that of the Syndrome. We believe that special search strategies for the decision tree method during the decision-making process could improve the decision tree results. However, this is left for further research.

The accuracy of peripheral lung cancer diagnosis with the use of the developed decision rules is close to the diagnostic results of a highly qualified expert and can improve the number of correct diagnose for physicians of different qualifications.

The methodology, developed in this task, can be used for knowledge acquisition in other medical tasks, in particular, we applied it for mammogram diagnosis in our recent work [8].

Finally, we can say that picture archiving systems in combination with data mining methods open the possibility for advanced computer-assisted medical diagnosis. However, it will not give the expected result if pictures and expert's descriptions are not stored in a standard format in the archives (e.g. in PACS) for further analysis. Since standard vocabulary (8) and very good experts are available for many medical diagnostic tasks, data mining methods can be used for mining knowledge that could support image analysis and classification in the case of uncertainty. If the vocabulary is not a priori available, it can be elaborated using, for example, repertory grid methodology, described in [10].

**Acknowledgements**

This work was possible thanks to the financial support from Russian Foundation for Basic Researches (Grant of RFFI N 99-01-00268.) and Grant of INTAS (Project 96-0833).

## References

1. In: "Proceedings of 15[th] Symposium for Computer Applications in Radiology: Filmless radiology –reengineering the practice of Radiology for the 21[st] Century". Baltimore, USA 1998. J. of Digital Imaging 1999; 11: 3, Suppl. 1.
2. Jorwekar G.J., Dandekar K.N. Baviskar P.K., Picture Archiving and Communication System (PACS): Clinician's Perspective About Filmless Imaging, Indian Journal of Surgery December 2015, Volume 77, Supplement 3, pp 774–777.
3. Benjamin Schooley; Neset Hikmet; Emrah Atilgan, Health IT Maturity and Hospital Quality: Effects of PACS Automation and Integration Levels on U.S. Hospital Performance, In Proc.: 16 International Conference on Computational Science and Computational Intelligence, 15-17 Dec. 2016, Las Vegas USA, DOI: 10.1109/CSCI.2016.0016
4. Cheryl A. Petersilge, The Evolution of Enterprise Imaging and the Role of the Radiologist in the New World, American Journal of Roentgenology, Volume 209, Issue 4, 209: 845-848. 10.2214/AJR.17.17949
5. Christian J. Herold, Jonathan S. Lewin, Andreas G. Wibmer, James H. Thrall, Gabriel P. Krestin, Adrian K. Dixon, Stefan O. Schoenberg, Rena J. Geckle, Ada Muellner, Hedvig Hricak, Imaging in the Age of Precision Medicine: Summary of the Proceedings of the 10th Biannual Symposium of the International Society for Strategic Studies in Radiology, RadiologyVol. 279, No. 1, https://doi.org/10.1148/radiol.2015150709
6. In:"Proceedings of SPIE International Symposium Medical Imaging 2000", San-Diego, USA, 2000, SPIE 3981.
7. IPCAI 2018 Special Issue: Information Processing for Computer-Assisted Interventions, 9th International Conference 2018—Part 1, Pascal Fallavollita, Etsuko Kobayashi, Stefanie Speidel (Eds.) International Journal of Computer Assisted Radiology and Surgery May 2018, Volume 13, Issue 5, pp 607–610
8. Heywang-Köbrunner S, Perner P. Optimized Computer-Assisted Diagnosis based on Data Mining, Expert Knowledge and Histological Verification, IBaI Report ISSN 1431-2360, 2000.
9. Perner P. and Attig A., Prototype-Based Classification for Automatic Knowledge Acquisition of Pathological Processes at the Cellular Level, Transactions on Mass-Data Analysis of Images and Signals 2(1), September 2010
10. Boose JH, Shema DB, Bradshaw JM. Recent progress in Aquinas: a knowledge acquisition workbench. Knowledge Acquisition 1989; 1: 185-214.
11. Perner P., Big data, Decision Tree Induction, and Image Analysis for the Discovery of Decision Rules for Colon Examination,
Journal on Tissue Science and Engineering December 2016, Volume 7, Issue3, ISSN:2157-7552, (suppl.), http://dx.doi.org/104172/2157 7552.C1.026, page 34
12. Perner P., Quantitative measurement of cellular events with image processing and data mining for drug discovery, therapy, and system biology, Journal of Proteomics Bioinform, 9:11, http.//dx.doi. org/10.4172/0974-276X.C1:090

13. Case-Based Reasoning for Image and Signals, P. Perner (Eds.), Series Computational Intellignece, Springer Verlag 2007
14. Perner P., Case-Based Reasoning for Image Analysis and Interpretation In: C. Chen and P. S. P. Wang, Handbook on Pattern Recognition and Computer Vision, 3rd Edition, World Scientific Publisher, 2005, pp. 95-114
15. Belikova T.P. Synthesis of optimal linear filters for visualization of diagnostically important objects in medical radiological tasks. In: "Digital Optics in Medical Introscopy", Moscow: Science, 1992: 57-72 (in Russian).
16. Barrett HH, Swindell W. Radiological Imaging: theory of image formation, detection and processing. San Diego: Academic Press, 1996.
17. Belikova. TP, Yashunskaya NI, Kogan EA. Computer-Aided differential Diagnosis of Small solitary Pulmonary Nodules. Computer and Biomedical Research 1996, 29: 1, 48-62.
18. Belikova TP, Yashunskaya NI, Kogan EA. Computer analysis for differential diagnosis of small pulmonary nodules. In: Proc. of Int. Congress for lung cancer. Monduzzi, editor, Athens, 1994: 93-98.
19. Perner P, Belikova TP, Yashunskaya NI. Knowledge Acquisition by symbolic decision tree induction for interpretation of digital images in radiology, In: Perner P., Wang P., Rosenfeld A., editors, Advances in Structural and Syntactical Pattern Recognition. Heidelberg: Springer-Verlag, lnai 1121, 1996: 208-219.
20. Data Mining Tool Decision Master, Description available at http://www.ibai-solutions.de
21. Pereverzev-Orlov VS. A Partner System and the Concept of Recognition Learning. Pattern Recognition and Image Analysis 1992, **2**: 420-437.
22. Perner, P., Decision Tree Induction Methods and Their Application to Big Data, In: Fatos Xhafa, Leonard Barolli, Admir Barolli, Petraq Papajorgji (Eds.), Modeling and Optimization in Science and Technologies, Modeling and Processing for Next-Generation Big-Data Technologies With Applications and Case Studies, Volume 4 2015, Springer Verlag 2015, pp. 57-88.
23. Weiss SM, Kulikowski C. Computer Systems that learn. Morgan Kaufmann, 1991.
24. Perner P. Data Mining for Multimedia Data. Heidelberg, Springer Verlag 2001.
25. Baird HS, Mallows CL. Bounded-Error in Pre-classification Trees. In: Dori D, Bruckstein A, editors, Shape, Structure and Pattern Recognition. World Scientific Publishing Co., 1995: 100-110.
26. Quinlain JR. Simplifying decision tree. Intern. Journal on Man-Machine Studies 1987, **27**: 221-234.
27. Perner P. Mining Knowledge in Medical Image Databases, In: Dasarathy, BV, editor, Data Mining and Knowledge Discovery: Theory, Tools, and Technology. 2000, SPIE 4057:359-369.
28. Belikova TP, Stenina I I, Yashunskaya N I. Computer-aided methods to recover strategies for visual search and navigation. In: SPIE International Symposium Medical Imaging 2000. SPIE 3981: 240-250:
29. Shapiro A. Structured Induction. John Wiley Inc., 1985.
30. Agresti A. Categorical Data Analysis. John Wiley Inc., 1990.
31. Jain AK, Dubes R. Algorithms for Clustering Data. Prentice Hall: Englewood Cliffs, 1988.
32. Belikova TP, Stenina I.I., Yashunskaya N.I. Image interpretation based on image processing and knowledge-guided data analysis. J. Machine Graphics & Vision 1998, 17: 1/2, 187-196.
33. Upton G. The Analysis of Cross-tabulated Data. J. Wiley & Sons, 1978.
34. Utgoff PE. Incremental Induction of Decision Trees. Machine Learning 1989, 4: 161-186.

**Table 1.** Classes and Attribute List (1), developed for processed tomograms with small nodules

| Attr. No. | Attribute Type | Attribute Name | Short name | Value N | Attribute Value |
|---|---|---|---|---|---|
| 1 | Boolean | Class | Class | 1 | Malignant |
| | | | | 2 | Benign |
| 2 | Categorical | Structure inside the module | StrInsNod | 1 | Inhomogeneous with disorderly structures |
| | | | | 2 | Inhomogeneous with orderly structure |
| | | | | 3 | Regularly decreasing film density along the periphery of the nodule |
| | | | | | Areas with calcifications |
| | | | | 4 | Enough homogeneous structures |
| | | | | 5 | Inhomogeneous with calcifications |
| | | | | 6 | Inhomogeneous with orderly structures and calcifications |
| | | | | 7 | Inhomogeneous with cavities |
| | | | | | |
| | | | | 8 | Inhomogeneous with cavities |
| 3 | Categorical | Scar-like changes inside the nodule | ScrLikeChan | 1 | Irregularly shaped fragmentary dense shadow |
| | | | | 2 | Regular dense shadow along periphery |
| | | | | 3 | None |
| 4 | Categorical | Shape | Shape | 1 | Nonround |
| | | | | 2 | Round |
| | | | | 3 | Oval |
| | | | | 4 | Lobular |
| | | | | 5 | Angular |
| 5 | Categorical | Margin | Margin | 1 | Nonsharp |
| | | | | 2 | Sharp |
| | | | | 3 | Nonsmooth |
| | | | | 4 | Smooth |
| | | | | 5 | Lobular |
| | | | | 6 | Angular |

| | | | | | |
|---|---|---|---|---|---|
| | | | | 7 | Spicular |
| | | | | 8 | Nonsharp-sharp (nonsharp in some regions and sharp in others) |
| | | | | 9 | Nonsharp and Nonsmooth |
| | | | | 10 | Nonsharp and Angular |
| | | | | 11 | Nonsharp and Spicular |
| | | | | 12 | Sharp and Smooth |
| | | | | 13 | Sharp and Lobular |
| | | | | 14 | Sharp and Angular |
| | | | | 15 | Nonsharp-sharp and Angular |
| 6 | Categorical | Convergence of vessels | ConvofVes | 1 | Vessels constantly exist converging to the nodule |
| | | | | 2 | Vessels are forced away the nodule |
| 7 | Categorical | Outgoing shadows in surrounding tissues | OutgoShadinTis | 1 | Chiefly vascular |
| | | | | 2 | Outgoing sharp tapelines (septa) |
| | | | | 3 | None |
| | | | | 4 | Invasion into surrounding tissues |
| 8 | Continuos | Size of Nodules | Size | | Values in cm (e. g. 1,2 := 1,2 cm) |
| 9 | Categorical | Character of the Lung pleura | CharPleu | 1 | Thickening |
| | | | | 2 | Withdrawing |
| | | | | 3 | None |
| | | | | 4 | Thickening with Withdrawing |

**Table 1.** Classes and Attribute List (1), developed for processed tomograms with small nodules (to be continued

**Table 2.** Classes and Attribute List (2), developed for processed tomograms with small nodules

| No. of Attribute | Attribute type | Attribute Name | Short Attribute Name | No. of Values | Attribute Value |
|---|---|---|---|---|---|
| ... | ... | ... | ... | ... | ... |
| 8 | Categorical | Sharpness of Margin | SharpMar | 1<br>2<br>3 | Nonsharp<br>Mixedsharp<br>Sharp |
| 9 | Categorical | Smoothness of Margin | SmoothMar | 1<br>2<br>3 | Nonsmooth<br>Mixedsmooth<br>Smooth |
| 10 | Boolean | Lobularity of Margin | LobMar | 0<br>1 | Nonlobular<br>Lobular |
| 11 | Boolean | Angularity of Margin | AngMar | 0<br>1 | Nonangular<br>Angular |
| 12 | Boolean | Spicularity of Margin | SpicMar | 0<br>1 | Nonspicular<br>Spicular |
| ... | ... | ... | ... | ... | ... |
| 14 | Boolean | Vascular Outgoing Shadow | Vascshad | 0<br>1 | None<br>Chiefly vascular shadows |
| 15 | Boolean | Outgoing Sharp thin tape lines | OutgoSha | 0<br>1 | None<br>Outgoing sharp thin tape lines |
| ... | ... | ... | ... | ... | ... |
| 18 | Boolean | Thickening of lung pleura | ThLungPl | 0<br>1 | None<br>Thickening |
| 19 | Boolean | Withdrawing of lung pleura | WithLupl | 0<br>1 | None<br>Withdrawing |
| ... | ... | ... | ... | ... | ... |

**Table 7.** List of features used for decision rule creation ("V" marks the feature used in Syndrome).

| No. of Attribute | Attribute type | Attribute Name | Short Attribute Name | No. of Values | Attribute Value | $m_1$ | $m_2$ |
|---|---|---|---|---|---|---|---|
| ***CLASS*** | Boolean | Class | Class | 1<br>2 | Malignant<br>Benign | **classifying feature** | |
| 1 | Categorical | Structure inside the module | StrInsNod | 1<br>2 | Irregular<br>Regular orderly | *V* | - |
| 2 | Categorical | Scar-like changes inside the nodule | ScrLikeChan | 0<br>1<br>2 | None<br>May be exist<br>Irregularly fragmentary dense shadow | *V* | *V* |
| 3 | Categorical | Shape | Shape | 1<br>2<br>3 | Nonround<br>Round<br>Oval | *V* | *V* |
| 4 | Categorical | Sharpness of Margin | SharpMar | 1<br>2<br>3 | Nonsharp<br>Mixedsharp<br>Sharp | *V* | - |
| 5 | Categorical | Smoothness of Margin | SmoothMar | 1<br>2<br>3 | Nonsmooth<br>Mixedsmooth<br>Smooth | *V* | *V* |
| 6 | Boolean | Angularity of Margin | AngMar | 0<br>1 | Nonangular<br>Angular | *V* | *V* |
| 7 | Boolean | Spicularity of Margin | SpicMar | 0<br>1 | Nonspicular<br>Spicular | *V* | *V* |
| 8 | Categorical | Convergence of vessels | ConvofVes | 1<br>2 | Vessels are forced away the nodule<br>Vessels are converging to the nodule | *V* | *V* |

**Table 7.** List of features used for decision rule creation ("V" marks the feature used in Syndrome) to be continued

| **No. of Attribute** | **Attribute Type** | **Attribute Name** | **Short Attribute Name** | **No. of Values** | **Attribute Value** | $m_1$ | $m_2$ |
|---|---|---|---|---|---|---|---|
| 10 | Boolean | Outgoing shadows: Invasion into surrounding tissues | OutgoShadinTis | 0<br>1 | None<br>Invasion into surrounding tissues | *V* | *V* |
| 11 | Boolean | Thickening of lung pleura | ThLungPl | 0<br>1 | None<br>Thickening | *V* | - |
| 12 | Boolean | Withdrawing of lung pleura | WithLupl | 0<br>1 | None<br>Withdrawing | *V* | - |

# Determination of Alzheimer Disease by the Calculation of the Brain/Liquor Quotient

Petra Perner

Institute of Computer Vision and Applied Computer Sciences
PF 30 11 14, 04251 Leipzig, Germany
e-mail: pperner@ibai-institut.de

**Abstract.** Degenerative brain diseases such as Alzheimer are increasing because of aging of people. To determine the disease in its earliest stage is very important. A measure for degenerative brain diseases is the brain/Liquor coefficient. Nowadays, this measure can be only manually detected in CT- or MRI-images. This subjective measure is very inexact. An objective measurement that allow to determine the measure at any time and compare the values of the measure between different time point would be preferable. In that way can be detected the disease and, also the progress of the treatment can be observed. Therefore, we need an automatic system that can acquire the CT- or MRI images of the skull of the patient and automatically analyses the images and determines the brain/liquor quotient. For such a system, image segmentation is a crucial step in extracting information from a digital image. It is not easy to set up the segmentation parameter so that it fits best over the entire set of images, which need to be segmented. In the paper, we propose a novel method for image segmentation based on CBR. We describe the whole architecture and the methods used for the various components of the systems and show how it performs on medical images. The evaluation is done based on human labelled images. We compare the with the automatic labeled images and can show superior results.

**Keywords:** degenerative brain diseases, Alzheimer, image analysis, brain/liquor measurement, CBR-based image segmentation

## 1 Introduction

Degenerative brain diseases such as Alzheimer are increasing because of aging of people. To determine the disease in its earliest stage is very important. A measure for degenerative brain diseases is the brain/Liquor coefficient [6]. Nowadays, this measure can be only manually detected in CT- or MRI-images. This subjective measure is very inexact. An objective measurement that allow to determine the measure at any time and compare the values of the measure between different time point would be preferable. In that way can be detected the disease and, also the progress of the treatment can be observed.

Image Segmentation is a crucial step in extracting information from a digital image. It is not easy to set up the segmentation parameter so that it fits best over the entire set of images. Most segmentation techniques contain numerous control parameters, which must be adjusted to obtain optimal segmentation performance. The parameter selection is usually done on a large enough test data set, which should represent the entire domain well enough in order to be able to built-up a general model for the segmentation. However, it is often not possible to obtain a large enough data set and therefore the segmentation model doesn't fit well to the entire data and needs to be adjusted to new data. Note that, a general model doesn't guarantee the best segmentation for each image rather than it guarantees an average best fit over the entire set of images.

Another aspect goes along with changes in image quality caused by variations in environmental conditions, image devices, etc. Thus, the segmentation performance needs to be adapted to these changes in image quality. All this suggests using CBR for image segmentation. We have successfully used CBR framework for the high-level unit of an image interpretation system [1]-[3] and can show the extra-ordinary good performance of this approach for image interpretation compared to other approaches. Grimnes et al. [4] used CBR for medical image understanding based on a propose-verify-critique-modify framework which expands the idea of hypothesis-test based systems like Ernest [5] by a learning component.

In the paper, we propose a novel image segmentation scheme based on case-based reasoning. Our approach doesn´t relay on a propose-verify framework. CBR is used to select the segmentation parameter according to the current image characteristics. By taking into account the non-image and the image information we break down our complex solution space to a subspace of relevant cases where the variation in image quality between the cases is limited. It is assumed that images having similar image characteristics will show similar good segmentation results when the same segmentation parameters are applied to these images.

We evaluate our method on a set of medical images (CT scans of the brain) where the variations between the images and objects in the images are naturally very high. The complexity of the brain CT-scans is due to partial volume effects, which disturb the edges and produce contrast degradation by spatial averaging, and to the typical problems such as patient movements, beam hardening, and reconstruction artifacts. These image characteristics are responsible for the over- and undersegemented results observed when unsupervised segmentation is applied.

We use the proposed method for labeling brain and liquor areas in CT slices. Based on this we calculate brain/liquor ratio which is a parameter to determine the degree of degenerative brain diseases [6].

In Section 2 of the paper, we describe the overall architecture. The case description is presented in Section 3. The segmentation algorithm is described in Section 4. The similarity measure for the non-image information and the image information is described in Section 5. Results are given in Section 6. Finally, we give conclusions in Section 7.

## 2 Overall Architecture

We will divide the overall architecture into the image segmentation unit based on case-based reasoning (see Fig. 1) and in the unit for the casebase management part (see Fig.2).

### 2.1 Case-Based Reasoning Unit

The case-based reasoning unit for image segmentation consists of a case base in which formerly processed cases are stored. A case comprises of image information, non-image information (e.g. image acquisition parameters, object characteristics and so on), and image segmentation parameters. The task is now to find the best segmentation for the current image by looking up the case base for similar cases. Similarity determination is done based on non-image information and image information. The evaluation unit will take the case with the highest similarity score for further processing. In case there are two or more cases with the same similarity score the first appeared case will be taken. After the closest case has been chosen, the image segmentation parameters associated with the selected case will be given to the image segmentation unit and the current image will be segmented, (see Fig. 1). It is assumed that images having similar image characteristics will show similar good segmentation results when the same segmentation parameters are applied to these images.

In the following we will assume for segmentation: the definition of regions based on constant local image features, and for labeling: the classification of regions into the object classes: brain and liquor.

In the approach for brain/liquor determination, the volume data of one patient (a sequence of max. 30 CT-image slices) are given to the CBR image segmentation unit. The CT images are stored in DICOM-format. Each file consists of a header and the image matrix. In the header are stored information about the patient and the image acquisition.

The images are processed slice by slice before the brain/liquor volume ratio can be calculated. First, each image is preprocessed in order to eliminate the non-interesting image details like the skull and the head shell from the image. Afterwards, from the image file header is extracted the non-image information (see Sect. 3.1). From the image matrix contained in the DICOM-file is processed the statistical features describing the image characteristics (see Sect. 3.2). This information together with the non-image information is given to the unit, which determines the similarity. The similarity between the non-image information and the image information of the current case and the cases in case base is calculated (see Sect. 5). The closest case is selected and the segmentation parameters (see Sect. 4) are given to the segmentation unit. The segmentation unit takes the parameters, adjusts the segmentation algorithm and segments the image into brain and liquor areas. The resulting liquor area is displayed on screen to the user by red coloring of the area in the original image. This is done in order to give the user a visual control of the result.

These processing steps are done slice by slice. After each slice has been processed, the volume for brain and liquor is calculated. Finally, the brain/liquor volume ratio is processed and displayed to the user.

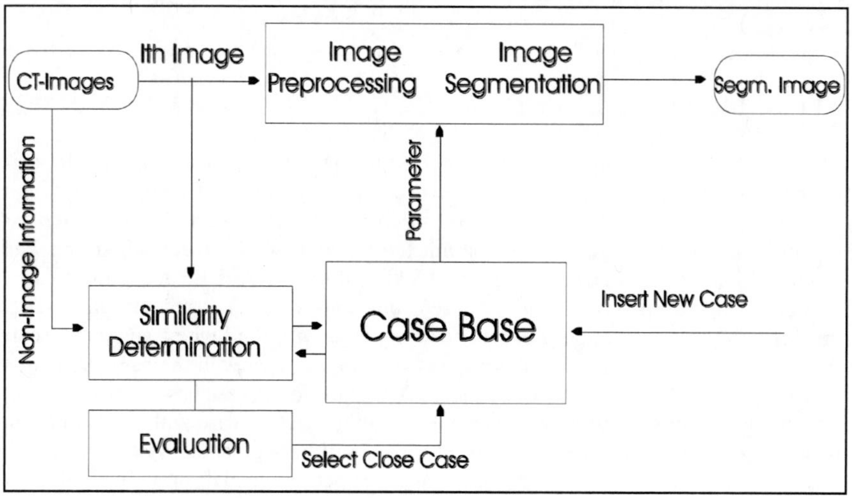


**Fig. 1.** Case-Based Reasoning Unit

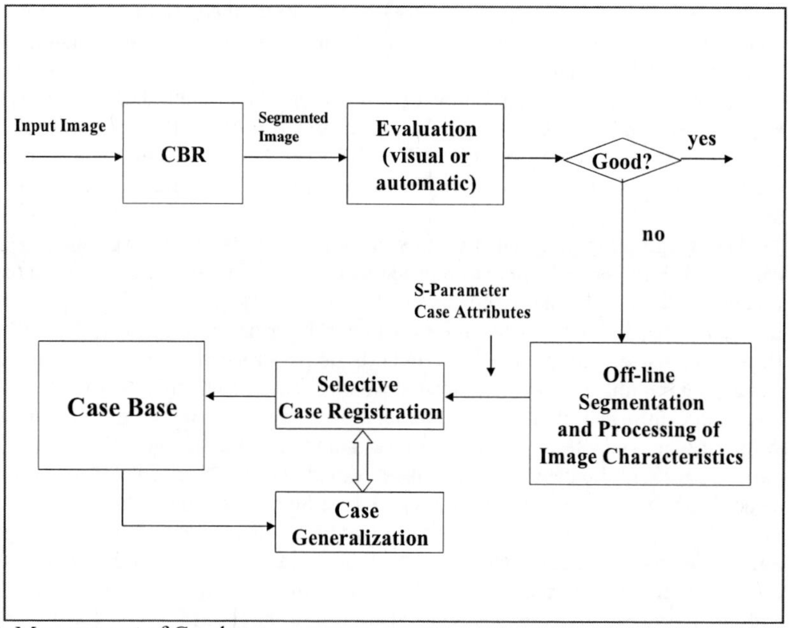


**Fig. 2.** Management of Casebase

## 2.2 Management of Casebase

The result of the segmentation process is observed by a user. He compares the original image with the labeled image on display. If he detects deviations of the marked areas in the segmented image from the object area in the original image, which should be labeled, then he will evaluate the result as incorrect and case-base management will start. This will also be done if no similar case is available in the case base. The proposed method is close to the critique-modify framework described by Grimnes et al. [4].

The evaluation procedure can also be done automatically [7]. However, the drawback is that there is no general procedure available. It can only be done in a domain dependent fashion. Therefore, an automatic evaluation procedure would constrain the usage of the system.

Once the user observes a bad result, he will tag the case as bad case. The tag describes the users critique in more detail. For the brain/liquor application we need to know the following information for the modification phase: too much or too less brain area, too much or to less liquor area and similarity value less than a predefined value. In an off-line phase, the best segmentation parameters for the image are determined and the attributes, which are necessary for similarity determination, are calculated from the image. Both, the segmentation parameters and the attributes calculated from the image, are stored into the casebase as new case. In addition to that, the non-image information is extracted from the file header and stored together with the other information in case base. During storage, case generalization will be done to ensure that the casebase will not become too large. In the current system, this function is not fully realized yet. It has to be done manually by the knowledge engineer. The unit for modifying the segmentation is shown in Figure 3.

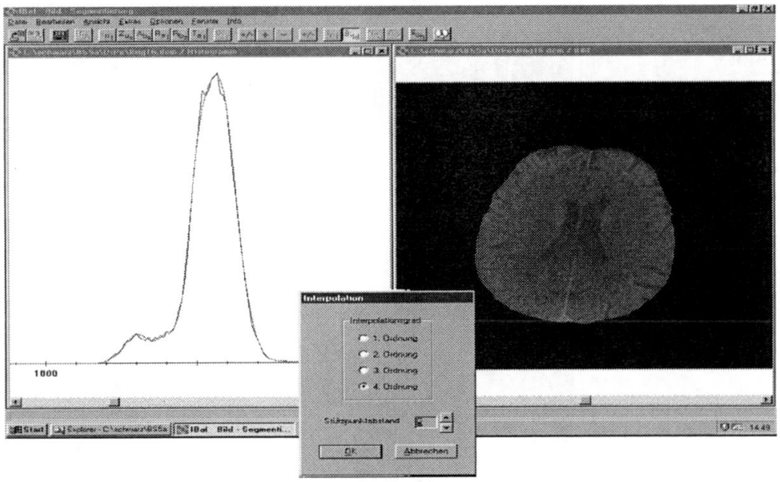


**Fig. 3.** User Interface of the Modification Unit

## 3 Case Structure and Case Base

A case consists of non-image information, parameters describing the image characteristics itself, and the solution (the segmentation parameters).

### 3.1 Non-Image Information

We will describe the non-image information necessary for our brain/liquor application. If we deal with other applications, then other appropriate non-image information is contained in the case. For example, in motion analysis [8], we have camera position, relative movement of the camera and the object category itself as non-image information.

For our brain/liquor determination in CT-images, we use patient specific parameter (like age and sex), slice thickness and slice number. This information is contained in the header of the CT image file so that we can automatically access these parameters.

Young Patients have smaller liquor areas than old patient. The images show different image characteristics. The anatomical structures and therefore the image characteristics also differ by women and men.

The slice number may vary from patient to patient because of the biological diversity and so may vary the starting position of the slices. Therefore, we map the numeric value onto three intervals: bottom, middle and top slices. These intervals correspond to the sphere of different image characteristics (see Fig. 4). The intervals can easily be calculated by dividing the slice number trough three. The remaining uncertainty in position can be neglected.

| | | | |
|---|---|---|---|
| Slice Number | 1 | ... | 10 |
| Bottom | 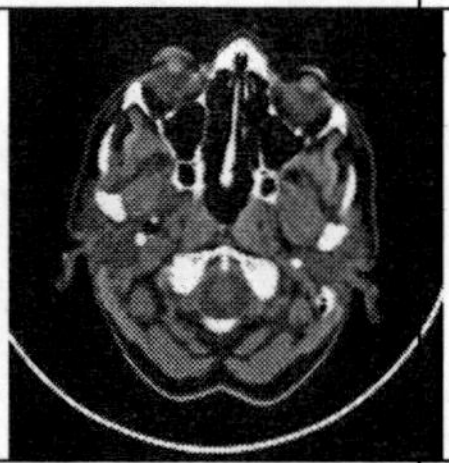 | ... | 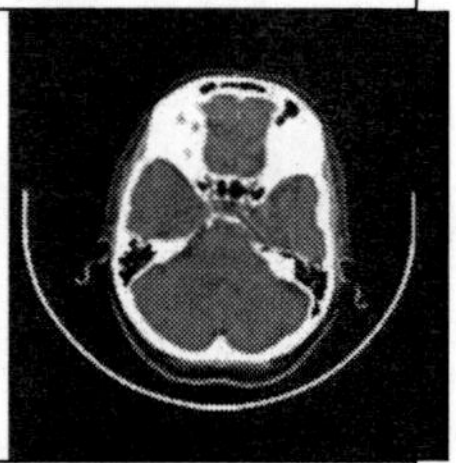 |
| Slice Number | 11 | ... | 21 |
| Middle | 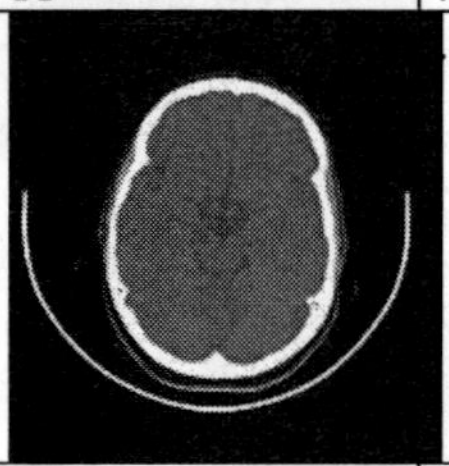 | ... | 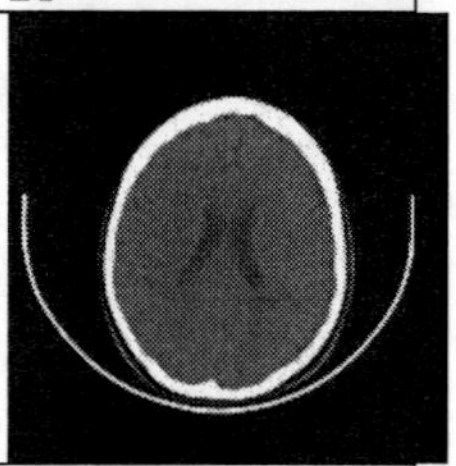 |
| Slice Number | 22 | ... | 31 |

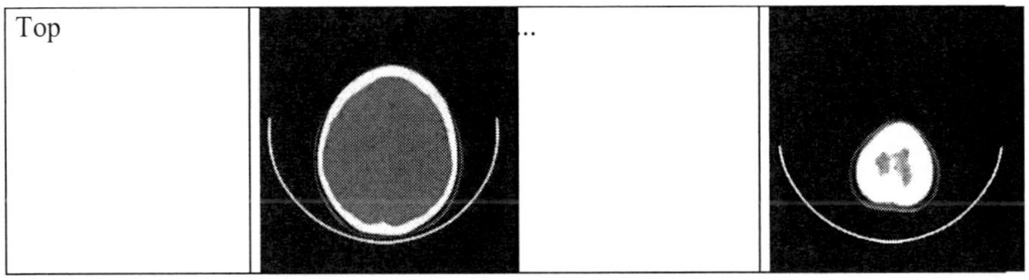


**Fig. 4.** CT Images showing the different Spheres of the Head

### 3.2 Image Information

The kind of image information describing a case is closely related to the kind of similarity measure used for similarity determination. There is a lot of work going on in developing new measures for comparing grey-scale images [9][10] for various objectives like image retrieval and image evaluation. Before we decided to take the similarity-measure described in that paper, we evaluated one of these new measures according to the used measure. Our decision for the particular similarity measure as well as the image information describing a case will be discussed here in short.

***IM1***

The similarity measure developed by Zamperoni et al [9] can take the image matrix itself and calculate from two image matrix the similarity (see Fig. 5). The input to the algorithm is the two images that should be compared. According to the specified distance function, the proximity matrix is calculated for one pixel at position r,s in image *A* to the pixel at the same position in image *B* and to surrounding pixel within a predefined window. The same is done for the pixel at position r,s in image *B*. Then, clustering is performed based on that matrix in order to get the minimum distance between the compared pixel. Afterwards, the average of both values is calculated. This repeats until all pixels of both images are processed. From the average minimal pixel distance, the final dissimilarity for the whole image is calculated.

The introduced window should make this measure invariant to scaling, rotation and translation dependent on the window size.

For that kind of similarity determination, we have to store the whole image matrix as image-information in a case. However, the similarity measure based on Zamperoni et al. has some drawbacks, which will be discussed later.

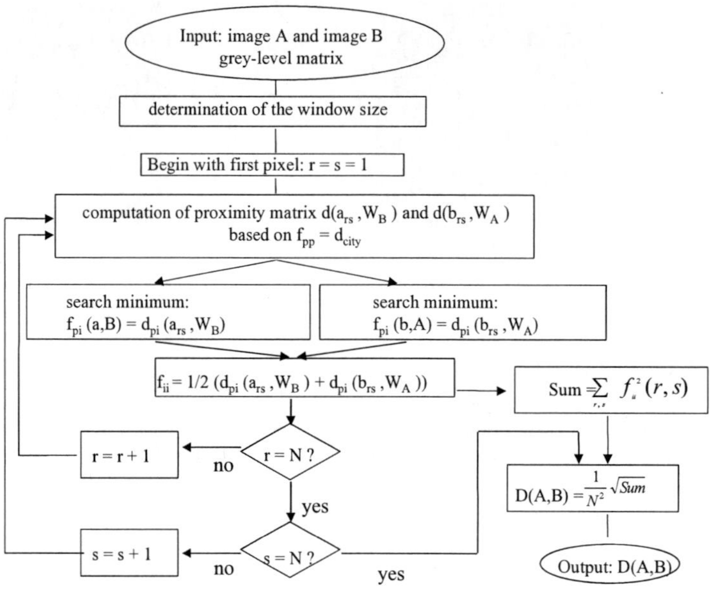


**Fig. 5.** Similarity Determination by Zamperoni et al

***IM2***

A simpler approach is to calculate features from the image, which describe the statistical properties of an image. These features are statistical measures of the gray level like: mean, variance, skewness, kurtosis, variation coefficient, energy, entropy, and centroid [11] (see Table 1). The similarity between two images is calculated based on these features.

| | | | |
|---|---|---|---|
| Mean | $\bar{g} = \sum_g g \cdot H(g)$ | Variance | $\delta_g^2 = \sum_g (g-\bar{g})^2 H(g)$ |
| Skewness | $g_s = \frac{1}{\delta_g^3} \sum_g (g-\bar{g})^3 H(g)$ | Curtosis | $g_k = \frac{1}{\delta_g^4} \sum_g (g-\bar{g})^4 H(g) - 3$ |
| Variation Coefficient | $v = \frac{\delta}{\bar{g}}$ | Entropy | $g_E = -\sum_g H(g) \log_2 [H(g)]$ |
| Centroid_x | $\bar{x} = \frac{\sum_x \sum_y x f(x,y)}{\sum_x \sum_y f(x,y)} = \frac{\sum_x \sum_y x f(x,y)}{\bar{g} S}$ | Centroid_y | $\bar{y} = \frac{\sum_x \sum_y y f(x,y)}{\sum_x \sum_y f(x,y)} = \frac{\sum_x \sum_y y f(x,y)}{\bar{g} S}$ |

| | |
|---|---|
| First order histogram | $H(g) = \frac{N(g)}{S}$ |

g - is the intensity value
N(g) - is the number of pixels of intensity value g in the image
S - is the overall number of pixels

**Table 1.** Image Features

***Comparison of IM1 and IM2***

We did investigations about the behavior of the two similarity measures, which are described in more detail in [12]. The similarity between the image slices of one patient are calculated based on IM1 and IM2. Single-linkage method is used to create the dendrogram, which graphically shows us the similarity relation between the slices. The dendrogram (see Fig. 6) based on IM2 shows two clusters: one for the slices at the middle and the top of the head and one for the slices at the bottom of the head. There is no clear cluster for the top and the middle slices. Nevertheless, the differences in the similarity values are big enough to make a distinction between these slices. The highest dissimilarity is recognized between the slices from the bottom, which happens because of the high complexity of the image structures in that sphere.

The dendrogram based on IM1 shows a finer graduation between the various slices (see Fig. 7). It can also better distinguish between bottom, middle and top slices. However, if we compare slices from different patients, it shows some drawbacks, which are caused by rotation, scaling and translation. The invariant behavior of this measure is restricted to the window size. The compensation of these effects requires a large window size, which on the other hand causes high computation time (more than 3 minutes on a 4-node system based on Power PC 604 and a window size of 30x30 pixels). This makes this measure not suitable for our problem.

***Results***

The similarity measure based on IM1 is limited invariant to rotation, scaling and translation. Therefore, we decided to take the similarity measure based on IM2. Moreover, in case of IM1, we have to store the whole image matrix as case and calculate the similarity over the entire image matrix. The computational costs for the similarity calculation are very high and so would be the storage capacity. The lower sensitivity of IM2 to the different spheres of the brain can be reduced by introducing the slice sphere as non-image information discussed in Section 3.1.

For the similarity measure based on IM2, we have only to calculate features from the images before we can store the case in case base. This calculation is of low computational cost. Each image is described by statistical measures of the gray level like mean, variance, skewness, kurtosis, variation coefficient, energy, entropy, and centroid. This information together with the non-image information and segmentation parameters comprises a case.

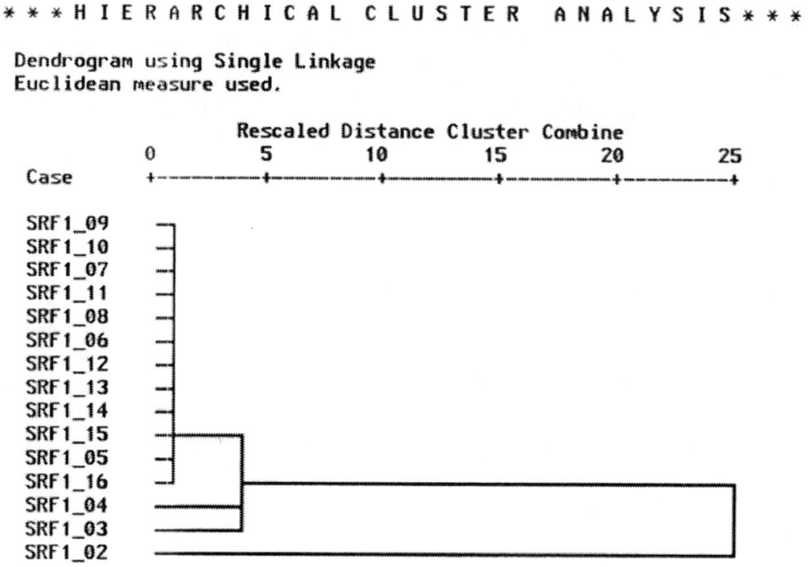


**Fig. 6.** Similarities between Images of one Patient based on IM1

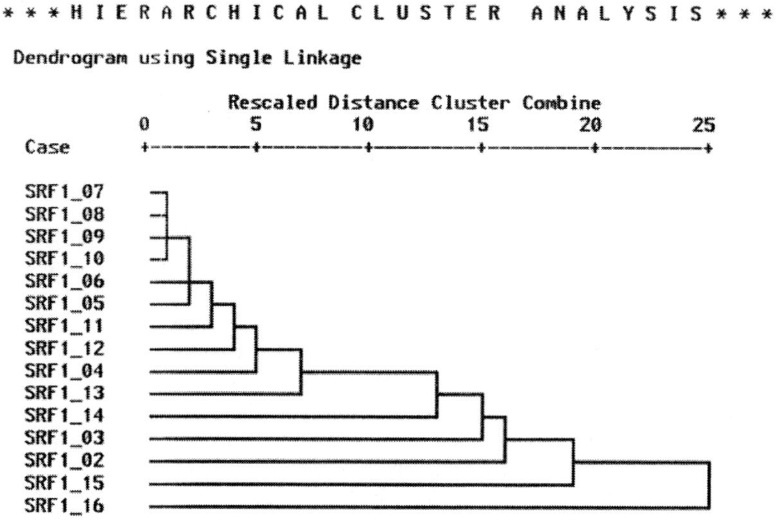


**Fig. 7.** Dendrogram of Similarities between Images of one Patient based on IM2

## 4. Segmentation Algorithm and Segmentation Parameters

The gray level histogram is calculated from the original image. This histogram is smoothed by some numerical functions and heuristic rules [13][14][15] to find the cut points for the liquor and brain gray level area. The parameters of the function and rules are stored with the cases and given to the segmentation unit if the associated case is selected.

The following steps are performed: The histogram is smoothed by a numerical function. There are two parameters to select: the complexity of the interpolation function and the interpolation width. Then the histogram is segmented into intervals such that each begins with a valley, contains a peak and ends with a valley. The peak to shoulder ratio of each interval is tested first: An interval is merged with the neighbor sharing the higher of its two shoulders if the ratio of peak height to the height of it's higher shoulder, is greater than or equal to a threshold. Finally, the number of the remaining intervals is compared to a predefined number of intervals. If more than this have survived, the intervals with the highest peaks are selected. The number of intervals depends on the number of classes the image should be segmented. The cut points are calculated and then applied to the image. Fig. 8 shows a histogram for an original image and the histogram after it was processed by the algorithm. The original image and the resulting labeled images are shown in Fig. 9.

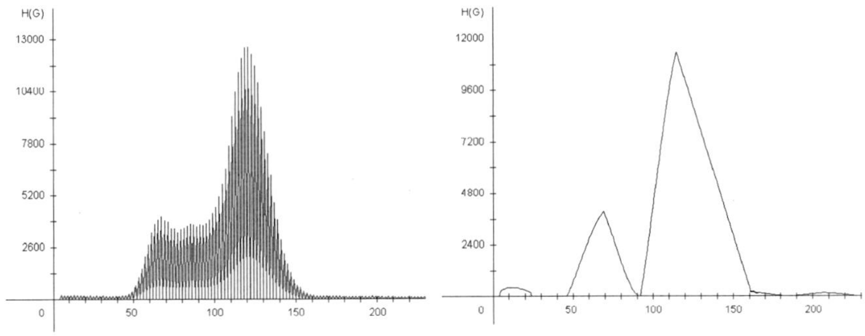


**Fig. 8.** Histogram of a CT-Image and refined Histogram

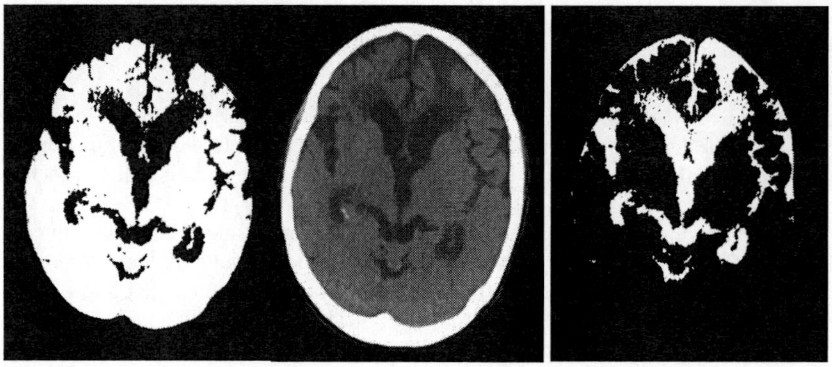

**Fig. 9.** Original Image, Labeled Images for Brain, and Labeled Image for Liquor

## 5 Similarity Determination

Similarity is comprised of two parts: non-image similarity and image similarity. The final similarity is calculated by:

$$Sim = \frac{1}{2}(Sim_N + Sim_I) = \frac{1}{2}\left(S(C_I, b) + 1 - dist_{AB}\right) \qquad \textbf{(1)}$$

We decided that non-image and image similarity should have equal influence to the final similarity. Only when both similarities have a high value, the final similarity will be high.

### 5.1 Similarity Measure for Non-Image Information

We use Tversky's similarity measure for the non-image information [16]. The similarity between a Case $C_i$ and a new case $b$ presented to the system is:

$$S(C_i, b) = \frac{|C_i|}{\alpha|A| + \beta|D| + \gamma|M|} \qquad \textbf{(2)}$$

$$\alpha = 1, \beta, \gamma = \frac{1}{2}$$

where $|C_i|$ is the set of attributes in case $C_i$, A is the set of corresponding attributes in case $C_i$ and *b*, *D* is the set of attributes having different values, and M is the set of attributes having missing values.

### 5.2 Similarity Measure for Image Information

For the numerical data, we use:

$$dist_{AB} = \frac{1}{k}\sum_{i=1}^{K}\left|\frac{C_{iA} - C_{i\min}}{C_{i\max} - C_{i\min}} - \frac{C_{iB} - C_{i\min}}{C_{i\max} - C_{i\min}}\right| \tag{3}$$

$C_{IA}$ and $C_{IB}$ are *i*th feature values of image A and B, respectively. $C_{iMIN}$ is the minimum value of the *i*th numeric or symbolic feature. $C_{iMAX}$ is the maximum value of the *i*th feature, and $W_i$ is the weight attached to the *i*th feature. For the first run, $W_i$ is set to one. Further studies will deal with learning of feature weights.

## 6 Results

***Evaluation***

The user interface of the recent system is shown in Fig. 10. Three slices of the sequence of slices for one patient are shown at the same time on display. In chronological order a new slice from this sequence is processed, the images at the display are scrolled from right to left and the actual processed image is displayed at the right side of the terminal. The original images are shown at the top of the display and the labeled images are shown at the bottom of the display, so that the user can also compare the images visually. The system is in practical use by the medical department at the university of Halle. The system contains 130 cases in casebase.

We compared the performance of the system with manual labeled images by the physicians and automatic labeled images by our system. Some of these images are contained in the casebase while others are not contained. In a BMFT study [17], one physician from the university of Leipzig labeled the images of 30 patients by hand. For each patient we have approx. 20 images that give a total of 600 images for evaluation. We also used images labeled by another physician from the university of Halle. This physician labeled the images twice so that we got an understanding of measurement error done by a human.

| Case | in CaseBase | Brain (ccm) | | Liquor (ccm) | | Brain/Liquor Ratio | |
|---|---|---|---|---|---|---|---|
| | | manual | autom. | manual | autom. | manual | autom. |
| | | | | | | | |
| NNA | contain | 1331,5 | 1197,89 | 287,76 | 316,98 | 4,63 | 3,78 |
| WFH | contain | 1211,2 | 1123,75 | 201,2 | 243,58 | 6,02 | 4,91 |
| MRG | contain | 1381,5 | 1078,2 | 314,51 | 346,13 | 4,39 | 3,12 |
| HNH | part_cont | 1097,3 | 1080,48 | 213,51 | 274,5 | 5,14 | 3,94 |
| MRR | part_cont | 1152,8 | 1212,08 | 144,92 | 145,01 | 7,95 | 8,36 |
| TRM | part_cont | 1420 | 1232,88 | 248,34 | 284,45 | 5,72 | 4,33 |
| MEI | not_cont | 983 | 986,31 | 147,95 | 180,27 | 6,64 | 5,47 |
| MNH | not_cont | 846,7 | 833,23 | 165 | 189,22 | 5,13 | 4,4 |
| .... | ... | ... | ... | ... | ... | ... | ... |

**Table 2.** Manual and Automatic Measures for Brain/Liquor for 9 Patients

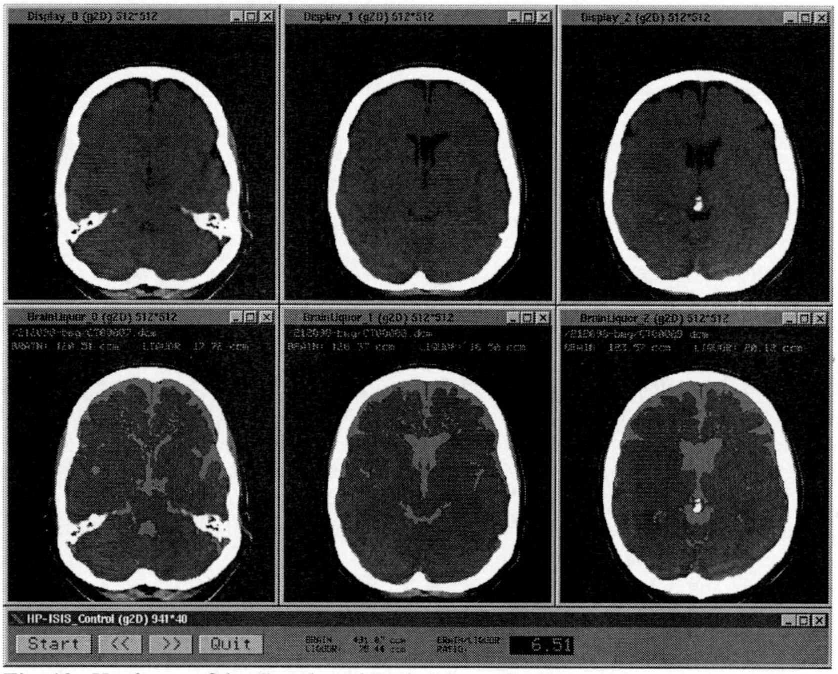


**Fig. 10.** Hardcopy of the Case-based Brain/Liquor System

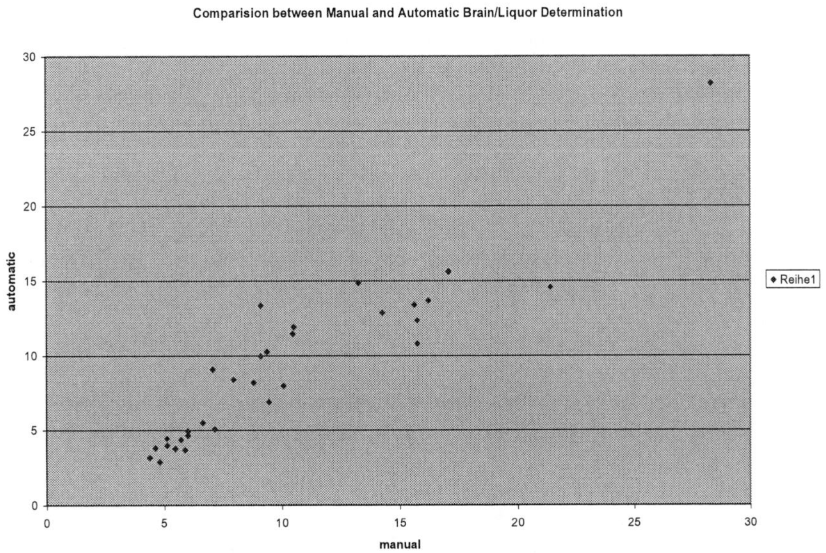


**Fig. 11.** Diagram Manual versus Automatic Results

Table 2 compares the manual results and the automatic results. Our algorithm labels more liquor area than a human expert does. However, Figure 11 shows a strong linear correlation between our results and the results of a human expert (r=0,85). This is a very good result.

***Knowledge Acquisition Aspect***

The case base has to be filled with a large enough set of cases. As described before, in the header of the DICOM-File are contained the non-image information. This information can be automatically extracted from the file and stored in case base. Likewise, the image information can be extracted from the image matrix contained in the DICOM-file. A little bit more effort is needed for the determination of the segmentation parameters. However, this task is efficiently supported by the acquisition unit shown in Fig. 3 The histogram is smoothed and processed step by step according to the implemented segmentation algorithm under the control of the knowledge engineer. The knowledge engineer can control each segmentation parameter and preview the segmentation result on screen. Once the best segmentation result has been reached the chosen segmentation parameters are stored together with the other information in case base.

***General Comments***

For the first time, a system automatically delivers to the physicians a measure for the brain/liquor ratio. The only thing that the physician has to do is to tell the system what images should be examined. No human interaction is necessary like for other volumetric image analysis systems [18]. The system automatically delivers a measure back to the physician. In contrast to the recent qualitative examination of the CT images, a quantitative examination is possible which gives a valid measure and allows a graduation between different stages of a disease and a control of a patient over time.

## 7 Conclusion

Degenerative brain diseases such as Alzheimer are increasing because of aging of people. To determine the disease in its earliest stage is very important. A measure for degenerative brain diseases is the brain/Liquor coefficient. Nowadays, this measure can only be manually detected in CT- or MRI-images. This subjective measure is very inexact. An objective measurement that allow to determine the measure at any time and compare the values of the measure between different time point would be preferable. In that way can be detected the disease and, also the progress of the treatment can be observed.

We have presented our methodology for a CBR based image segmentation system. The system performs image segmentation by looking up a case base for similarly processed images and takes the segmentation parameters associated to the similar image in order to do segmentation of the current image. We examined our system by comparing manually labeled images with automatic labeled images. The results show that by our method good results for the brain/liquor ratio can be obtained. Since it is well known that a physician doesn't always know what he should take as brain and liquor object we are not seriously concerned about the difference in labeling between a human and the system. A good evaluation of the system would only be possible if there would exists a true gold standard which might be available if the work on simulation of the brain CT-images will show good results.

With our system, we have given to the user a fully automatic system, which needs no user interaction when calculating the brain/liquor ratio. Such a system gives the opportunity to overcome the qualitative measure based on a subjective judgement to a quantitative measure, which is reproducible.

Further work will be done on generalizing cases and segmentation parameters and learning feature weights. Besides that, the system will be extended to other volumetrie-tasks like e.g. measurement of the volume of human liver and liver metastases.

## Acknowledgment

We like to thank Prof. Heywang-Köbrunner and Dr. Beck from the medical department of the university of Halle for the cooperation on this work. For providing us the results of the BMFT study, we like to thank Prof. Dietrich from the medical department of the university of Leipzig.. The project has been sponsored by the European Commission under the title "High-Performance Computing for Image Segmentation HPISIS".

## References

1. P. Perner, Different Learning Strategies in a Case-Based Reasoning System for Image Interpretation, Advances in Case-Based Reasoning, B. Smith and P. Cunningham (Eds.), LNAI 1488, Springer Verlag 1998, S. 251-261.
2. P. Perner, Case-Based Reasoning For Image Interpretation in Non-destructive Testing, 1st European Workshop on Case-Based Reasoning, Otzenhausen Nov. 1993, Proc. SFB 314 Univ. Kaiserslautern, Hrsg. M. Richter, vol. II, pp. 403-410
3. P. Perner, Using CBR Learning for the Low-Level and High-Level Unit of a Image Interpretation System, ICAPR´98, Plymouth, peer reviewed conference, In: Sameer Singh (Eds.), Advances in Pattern Recognition, Springer Verlag 1998, p. 45-54.
4. M. Grimnes and A. Amondt, A Two Layer Case-Based Reasoning Architecture for Medical Image Understanding, In: I. Smith and B. Faltings (Eds.), Advances in Case-Based Reasoning, Springer Verlag 1996, lnai 1168, pp. 164-178.
5. F. Kummert et al., Control and Explanation in a Signal Understanding Environment, Signal Processing 32 (1993), pp. 111-145.
6. Bettin, J. Dietrich, C. Dannenberg, H. Barthel, D. Zedlick, K. Jobst, W.H. Knapp, "Früherkennung von Hirnleistungstörungen – Vergleich linearer und volumetrischer Parameter (CT) mit Ergebnissen der Perfusions-SPET," 78. Deutscher Röntgenkongreß Wiesbaden 1997
7. S. Zhang, "Evaluation and Comparision of different Segmentation Algorithm," Pattern Recognition Letters, v. 18, No. 10, pp. 963-968, 1997.
8. G. Kummer and P. Perner, Motion Analysis, IBaI Report January 1999, ISSN 1431-2360
9. P. Zamperoni and V. Starovoitov, „How dissimilar are two gray-scale images", In Proc. of 17. DAGM Symposium 1995,Springer Verlag, pp.448-455
10. D.L.Wilson, A.J. Baddeley, and R.A. Owens, A new metric for grey-scale image comparision, Intern. Journal of Computer Vision, vol. 24, 1997, Issue 1, p. 1-29.
11. H. Dreyer and W. Sauer, Prozeßanalyse, Verlag Technik Berlin 1982
12. G. Kraft and P. Perner, Comparision of different Image Similarity Measures, IBaI Report Sept. 1998, ISSN 1431-2360
13. R. Ohlander, K. Price, and D.R. Reddy, "Picture Segmentation using recursive region splitting method," Comput. Graphics and Image Processing, 8: 313-333, 1978
14. C.H. Lee, "Recursive region splitting at the hierarchical scope views," Computer Vision Graphics, and Image Processing, 33, 237-259, 1986
15. P. Perner, Similarity-Based Image Segmentation, IBaI Report 1996 ISSN 1431-2360
16. A. Tversky, „Feature of Similarity", Psychological Review, vol. 84, No. 4, pp. 327-350, 1977.
17. Alzheimer Study "Degenerative Erkrankungen des zentralen und peripheren Nervensystems - Klinik und Grundlage", BMFT Study, Abschlußbericht der medizinische Fakultät der Uni Leipzig 1996
18. A. Tschammler et. al,"Computerized tomography volumetry of cerebrospinal fluid by semiautomatic contour recognition and gray value histogram analysis", Rofo Fortschr. Geb. Roentgenstr. Neue Bildgeb. Verfahren 1996, Jan: 164(1): 13-1

# MRI brain imagery processing software in data analysis

Alexander Bernstein[1], Akzhigitov Renat[1,2], Ekaterina Kondrateva[1], Svetlana Sushchinskaya[1], Irina Samotaeva[1,2,3], Vladislav Gaskin[1,2]

[1]Skolkovo Institute of Science and Technology, Moscow, Russia.
[2]Moscow Research and Clinical Center for Neuropsychiatry, Moscow, Russia.
[3]Laboratory of applied physiology of higher nervous activity,
Institute of higher nervous activity and neurophysiology, RAS, Moscow, Russia.

a.bernstein@skoltech.ru,
ekaterina.kondrateva@skoltech.ru,
svetlana.sushchinskaya@skolkovotech.ru

**Abstract.** Modern brain scanning techniques are one of the most consistent sources of medical information which included in most diagnostic protocols. These data possess a great potential for the machine learning (ML) analysis and clinical applications of the results. MRI imaging is prevalent in brain analysis, allowing acquisition of structural (3D image) and functional imaging (timeseries of 3D images). Relative to the MRI data volumes these research are considered to the big data and massive analysis, connected with numerous methods of signal processing, advanced ML approaches and feature extraction. In current paper we update the surveys of most software used for the feature generation, processing and analysis of MRI brain imagery. Systematizing the toolboxes, we describe common scanner characteristics, image modalities, corresponding features and their informational context, as well as common machine learning classification problems in working with the data. And as a result, we propose evaluation and comparison of conventional software tools with their detailed description.

**Keywords:** MRI Software Packages, MRI toolbox, Machine learning, Preprocessing, Classification, Neuroscience.

## 1 Introduction

Much progress has been made over the last years in computational neuroscience. The storage, visualization, processing and analysis of the medical imagery data are not feasible without computer-based techniques.

There is a constantly growing number of machine learning studies based on neuroimaging data. These research aimed to develop diagnostic and clinical prognostic prediction tools, facilitate an automatic segmentation or contribute in understanding the mechanisms of diseases by revealing new biomarkers and patterns.

A problem of machine learning classification in MRI imagery is usually related to multistage pipelines with handcrafted feature generation in different software packages, feature selection and analysis by methods of machine learning. Thus, the ML results are directly connected to the scanner characteristics, image modality and content, processing software and generated features quality.

Currently, there are various software packages, each targeted on different analysis and providing numerous features for users. Since there is no single package that can provide all of possible informative features, it is helpful to know the output of each software package, as well as the internal software characteristics [1].

In this paper, we provide the general information concerning different MRI modalities with related features, open databases and comparison of several software packages for feature generation, which are frequently used for the classification.

## 2 MRI data analysis

### 2.1 Imaging in neuroscience

The visualization in neuroscience could be achieved with usage of various scanning methods, covering Magnetic-Resonance Imaging (MRI), X-ray Computed Tomography (CT), Molecular imaging: Positron Emission Tomography (PET), Single-Photon Emission Computed Tomography (SPECT), Diffuse Optical Tomography (DOT), High-Resolution EEG (HR-EEG), Magnetoencephalography (MEG), Ultrasound imaging and their combinations [2].

Among of these methods the Magnetic-Resonance Imaging and Computed tomography are the most widely spread in clinical practice. The CT is beneficial in patients with implantable medical devices, such as cardiac pacemakers, vascular clips and cochlear implants as for these patients MRI scanning is contraindicated; in disturbed or very heavy patients, because the imaging can be performed much more rapidly, comparable to MRI.

In current paper we are focused on the MRI imagery because in most cases it is preferred in brain imaging due to the greater range of available tissue contrast, detailed anatomy and high sensitivity to abnormalities [3]. There is the superiority of MRI analysis in children and patients undergoing several imaging examinations explained by its potential harmlessness.

### 2.2 Medical Image Data and Datasets in terms of machine learning

The huge potential of machine learning approaches was proven in image-based diagnosis, risks evaluation and clinical prognosis [4], [5] .

Despite this, currently in clinical practice each patients scan is observed individually by a radiologist. For this approach the massive data analysis is not required: all imagery can be processed on one local machine (for visualization, morphometry estimation, etc.).

In contrast, the machine learning application to the brain imagery data is a massive data analysis. Then storing, managing and processing the 3D images with millions of voxels, or 4D sequences containing the timeseries for each voxel multiplied on the number of recordings and different acquisition modalities, became critical.

Using the machine learning one can process the original data in 3D arrays, but the most popular approach is to generate the vector of *features* consisted of hundred/thousands characteristics responsible for the specific brain property and proceed the data for the classification; most frequently these vectors are generated by the MRI brain analysis software.

It should be mentioned, that the size and quality of training dataset have the dominant influence on a ML model performance. Currently neuroscience research doesn't have large enough datasets (millions of samples) and the data collecting and sharing are critical in order to learn as much as we can about the neurological conditions [6].

There are still open questions concerning the collection, annotation, discovery, and sharing or reuse of medical imaging data.

### 2.3 Major MRI imaging properties

Existed MRI scanners provide various *spatial characteristics*, *magnetic field strengths* as well as different *imaging protocols*.

*Spatial characteristics*: major scanner features are represented by the spatial in-plane resolution, voxel size and a slice thickness. These characteristics mostly depend on the scanner inductive detection coil parameters [7].

*Magnetic Field:* Nowadays, there are scanners with from 1.5T to 7T magnetic field strength in clinical practice [8]. The higher magnetic strength field exposes higher signal-to-noise and contrast-to-noise ratio, higher resolution, enhanced contrast images which is beneficial in diverse pathologies, like multiple sclerosis, tumors, degenerative and age-related changes. Thus, additional information can be gained compared to imaging with lower field strength, in cause of better depiction of small anatomical details [9].

However, imaging at high field as 7T is tackled with several limitations as: the fields inhomogeneity, a higher specific absorption rate and additional contraindications for patient scanning and most importantly: higher field tomography is not widely spread [10].

In lower strength fields (1.5T vs 3T) the results of scanning may also differ. It was shown that the quality of 3T brain scan can be more feasible, but in most cases the informativity is comparable to 1.5T scan [11].

*Imaging protocols*: there are plenty of acquisition protocols and the number is evolving in parallel to hardware development and modernization of signal processing methods. The overall set could be divided to two broad categories: anatomical (structural) and functional scans.

In addition to the variety of protocols in different scanners, there is a possibility to manually adjust the protocol, for instance changing the echo time (TE), repetition time (TR) or switching the acquisition mode from axial to sagittal.

The characteristics listed above will directly affect the image quality and content; influence of these parameters on the processing results are still under the investigation. Thus, conducting the research, it is crucial to think of the dataset homogeneity and consider the level of the related uncertainty in case if the dataset contains imagery from different sources.

In this regard, there is a need to stick the research reporting and data managing protocols for the MRI/fMRI study and completely describe the used procedure and apparatus.

### 2.4 Structural MRI features

The structural imaging is aimed to provide the imagery for the static brain structure, enhancing desired patterns with usage of different acquisition modalities.

The anatomical scan types vary by acquisition time (AT), the time to echo (TE) and repetition time (TR) weighting: shorter TR relates to so-called T1 (spin-lattice) relaxation and affects contrast on T1-weighted images, longer TE relates to T2 (spin-spin) decay and affects contrast on T2-weighted images; fluid-attenuated inversion recovery (FLAIR) - by very long TR and TE times, and proton density weighting - by long TR and short TE, showing the difference in brain tissue physical characteristics. Structural scan can also be diffusional, or perfusion weighted (DWI, PWI), susceptibility weighted pointing the distortion of local magnetic field (SWI), specialized or enhanced (magnetization-prepared rapid gradient-echo (MP-RAGE), etc.).

The canonical ones for the structural imaging are: T1-weighted structural image (T1), T2-weighted fluid-attenuated inversion recovery structural image (T2 FLAIR), diffusion-weighted imaging (DWI), susceptibility-weighted imaging (SWI). Their usage in clinical practice is prescribed in most diagnostic protocols and it was proven that these 4 sequences are sufficient for the evaluation of patients with a new neurological complaint [12].

Features: sMRI images contain the anatomical (morphometric) features explaining the brain structural characteristics: gray matter surface area, cortical volumes thicknesses, sulcal depth, curvatures in sulcal regions, and sulcal walls. These data can be vectorized to the one-dimensional explanation of patients' cortex and subcortical nuclei (based on T1 or T2-weighted structural imagery).

Specifically patterns of calcification and hemorrhage could be enhanced by in the SWI imagery and then recognized more correctly by machine learning methods of segmentation.

White matter tracts (tractography) contains an information of so called structural connectivity reflecting which regions of interest (ROI) are structurally connected between each other. These data could be represented in graphs as well as in vectorized and then processed to the ML (DWI, DTI imagery)

The functional unit commonly used in brain imagery analysis is the region of interest: a subset of voxels identified for a specific purpose. These ROIs are arranged in different atlases parceling the brain tissue according to its known or hypostatized functionality [13].

### 2.5 Functional MRI features

The functional MRI investigation used for the study of the brain function based on measure of the brain neural activation reflected in blood oxygen-level dependent signal (BOLD signal). The fMRI study is not represented in most counties clinical practice, but in some – included in presurgical protocols [14]; otherwise conducted mostly for the research purposes.

For the functional imagery there are two types being frequently represented: resting-state functional MRI time series (rfMR) and task functional MRI time series data (tMRI). They could be observed in tandem or only rsMRI separately.

Most fMRI scans are gradient-echo spin-echo planar imaging (GE-EPI or SE-EPI) and therefore are particularly sensitive to artifacts in ROI typically affected by susceptibility effects.

*Features:* The fMRI series represented a series of compressed (in several folds lower resolution) 3D images. Popular functional connectivity analysis is described as a level of signal correlation between desired ROIs, highly correlated regions in BOLD are considered as functionally connected or involved in one neurocognitive process.

Thus, there is a possibility to observe 1D time series for desired ROIs, to conduct an analysis using the functional connectivity symmetrical matrixes or perform a graph-based analysis on these data.

Functional connectivity (fMRI) under the masking of structural connectivity (DWI or DTI) or hd-EEG and MEG will result in multimodal effective connectivity analysis [15].

The atlases from structural imaging could be applied to functional one considering its lower dimensionality.

## 3 Machine learning in MRI neuroimaging

As it was stated above, the machine learning is in a great request in many fields of study for biomedical tasks. The classical machine learning problems are classification and regression. As an input machine learning classifier get sample data and then, as an output give the probability of belonging to a particular class [16]–[22].

Modern machine learning techniques are quite successful for picture-based tasks. Recent Deep Convolutional Neural Networks were created under the biological inspiration and provide efficient data analysis [23]. Many diagnosis and disease prognosis objectives were explored and yielded new finding with modern data analysis techniques; for instance, in case of Alzheimer's Disease [24]–[28].

There are two main methods for picture-based tasks considering MRI and fMRI: to work with initial picture or to generate features from the images. If we do not generate features there is a problem with the interpretability of results.

Machine learning in neuroscience is quite complicated task because of high dimensionality of the data and relatively small dataset. For diagnostic aims it is easier to work with interpreted features. In this case we consider the vector representation of the brain with interpretable parameters *features*, discussed above. For instance, for the MRI such features are volumes, thicknesses, intensities, curvatures and surface areas of different brain regions and application of classical machine learning algorithms such as support vector machines (SVC), Random Forest (RFC), k nearest neighbors (kNC) or logistic regression (LR) used for classification.

The classical approach for machine learning task for identifying the patterns and trends is to obtain or generate features and then train classifiers on generated data. For this purpose, usually, external software toolboxes are used. As nowadays iPython is the most powerful instrument for data analysis it is more convenient use toolboxes implemented in iPython. One of the most popular project integrated to iPython is Nipy. It contains not only Analysis Pipeline Management for the most popular Neuroimaging tools, but also Nilearn - fast and easy way of statistical learning on neuroimaging data.

Since neuroimages have complex structure, there is a need for neuroimage based problems special ML approaches development. Often before analyzing neuroimages dimensionality reduction is needed, before application of classification approaches. Such methods of manifold learning as kernel principal component analysis(kPCA) are represented in most of the basic languages environments libraries (iPython, Matlab, etc.).

Another complication of neuroimaging tasks is overfitting. This happened because of small datasets for training and the sophisticated data structure. The overfitting might be controlled by cross-validation process - technique for estimating the performance of machine learning algorithms. The variety of cross-validation techniques are implemented in iPython (Sklearn lib) as well as in Matlab.

The most modern way of exploring initial data are deep learning algorithms. For the MRI or fMRI classification 3D or 4D Convolutional Neural Networks could be used. These algorithms could be used for large number of data. They provide powerful pipeline for the local feature generating of brain patterns and special techniques as batch normalization should be implemented to avoid overfitting.

## 4 MRI processing software

There are hundreds of available toolboxes/add-ints/tools for MRI data managing (BIDS), converting (Nibabel), visualization (3D Slicer), masking (MRIcron), atlasing (AAL), preprocessing (MRIQC, Mindboggle, etc.), preprocessing and analysis (SPM, FSL etc.), specialized analysis (Dype, CONN, LST, VLSM, etc.) predefined Pipeline resources (data, executable modules, workflows and services).

The features, observed in the proposed comparison, were chosen the way to give a newcomer to fMRI data analysis an overview of existed software packages potentials from the very general information to more detailed.

The list of packages with diverse usage and their basic information is given in the Table 1.

The medical significance of the toolbox is indirectly confirmed by the frequency of use of relevant articles; the reference papers for most popular open-source toolboxes and the number of their citations are represented in the Table 2.

The Table 3 contains the detailed information for the most popular software.

In the Table 4 we collated the description of the software/services providing pipelines for the several toolboxes.

The information in the tables is represented according to the following characteristics:

- URL address: the working and up-to-date URL supporting the software package;
- Use: the targeted purpose of the software; could be divided into;
- Availability: the information whether the package is an open source or commercially distributed;
- Referencing paper/Announcement date: the release date information, citing paper;
- Number of references: retrieved from Google Scholar search up to the 15 February 2018;
- Support: the supporting organization, foundation or university;
- Platform/operating system: software and hardware required for the installation;
- Preprocessing: consisted of the raw data obtained from the MRI scanner corrections, including artifact detection, spatial realignment, movement correction, centering, normalization, and smoothing in one pipeline that could be manually digested;
- Data input: the software availability to work with the 2D, 3D, and 4D images;
- Atlasing: working with the region of interest (ROI) input, as amount of default preinstalled atlases, the possibility of manual atlas editing or working with the custom one;
- Statistical analysis: if the package has internal module for the statistical analysis;
- Extensions: prescribed add-on extensions increasing the software functionality diversity and specificity.

**Table 1.** The List of the software for brain MRI data managing, processing and analysis. *V- visualization, P- preprocessing, A- analysis (including group analysis); O- open source, C- commercially distributed.*

| Package | Reference URL | Use | Access |
|---|---|---|---|
| AFNI | http://afni.nimh.nih.gov/afni/ | V,P,A | O |
| Analyze | http://www.analyzedirect.com/ | A | C |
| ANTS | http://www.picsl.upenn.edu/ANTS/ | V,P, A | O |
| BrainVoyager | http://www.brainvoyager.com/ | V, A | C |
| DCMTK | http://www.dcmtk.org/ | P | O |
| E-Prime | http://www.pstnet.com/eprime.cfm | A | C |
| fMRI Power | http://www.fmripower.org/ | A | O |
| Freesurfer | http://surfer.nmr.mgh.harvard.edu/ | V,P,A | O |

| | | | |
|---|---|---|---|
| FSL | http://www.fmrib.ox.ac.uk/fsl/ | V,P,A | O |
| ITK | http://www.itk.org/ | A | O |
| MRI Studio | https://www.mristudio.org/ | V | O |
| MRICron | http://www.cabiatl.com/mricro/mricro/mricro.html | P | O |
| Presentation | http://www.neurobs.com/ | P | C |
| SPM | http://www.fil.ion.ucl.ac.uk/spm/ | V,P, A | C |
| 3D Slicer | https://www.slicer.org/ | V,P | C |
| SpinalCord Toolbox | https://sourceforge.net/projects/spinalcordtoolbox/ | A | O |
| Nilearn | http://nilearn.github.io/ | A | O |
| BioimageSuite | http://bioimagesuite.yale.edu/ | A | O |
| BrainMagix | https://www.imagilys.com/brainmagix-clinical-neuroimaging-software-suite | P | C |
| FMRLAB | https://sccn.ucsd.edu/fmrlab/ | A | O |
| GraphVar | https://www.nitrc.org/projects/graphvar/ | A | O |
| ISAS | https://nonlinearestimation.bitbucket.io/ | P | O |
| AIR | http://www.cabi.gatech.edu/mricro/mricro/ | P | O |
| MRtrix3 | http://www.mrtrix.org/ | A | O |
| NIAK | http://niak.simexp-lab.org/ | P | O |
| Tortoise | https://www.nitrc.org/projects/tortoise/ | P | O |
| Humain Connectome Project | http://www.humanconnectomeproject.org/ | A | O |
| Ciftify(HCP) | https://github.com/edickie/ciftify | P | O |
| GIFT | http://mialab.mrn.org/software/gift/ | A | O |
| BART | https://mrirecon.github.io/bart/ | P | O |

| | | | |
|---|---|---|---|
| CAT12 | http://www.neuro.uni-jena.de/cat/ | P | O |
| CIVET | https://mcin-cnim.ca/technology/civet/ | P | O |
| Neuro3D | https://www.healthcare.siemens.com/ | P | C |
| NordicNeuroLab | http://www.nordicneurolab.com/ | P | O |

**Table 2.** Number of references to open-source software, obtained from the Google Scholar Search up to 15 February 2018.

| Toolbox | Software Name (Reference to the original paper) | # citations |
|---|---|---|
| SPM | Statistical parametric maps in functional imaging: A general linear approach(K.J. Frinston, 1995) | 8924 |
| AFNI | Software for analysis and Visualization of Functional Magnetic Resonance Neuroimages.(Robert W. Cox, 1996) | 6366 |
| FSL | Advance in functional and structural MR image analysis and implementation as FSL. (Stephen M. Smith, 2004) | 6987 |
| FreeSurfer | FreeSurfer: Whole brain segmentation: automated labeling of neuroanatomical structures in the human brain.(B. Fischel, 2002) | 3961 |
| AIR | Automated Image Registration: I. General Methods and Instrasubject, Intramodality Validation.(Roger P. Woods, 1998) | 1893 |
| MRICRO | Stereotaxi display of brain lesions. (C. Rorden, 2000) | 1701 |
| CONN | A functional connectivity toolbox for correlated and anticorrelated brain networks. (Whitfield-Gabrieli, S., and Nieto-Castanon, A. 2012) | 650 |
| Brain Voyager | Analysis of functional image analysis contest (FIAC) data with brainvoyager QX: From single-subject to cortically aligned group general linear model analysis and self-organizing group independent component analysis (R. Geobel, 2006) | 778 |
| Ciftify (HCP) | The minimal preprocessing pipelines for the Human Connectome Project. (Glasser MF, 2013) | 603 |
| VOXBO | Viewing facial expressions of pain engages cortical areas involved in the direct experience of pain. (M Botvinick, 2005) | 584 |
| BrainSuite | BrainSuite: An automated cortical surface identification tool | 564 |
| PLS | Partial least squares analysis of neuroimaging data applications and advances.(A. R. McIntosh,2004) | 561 |
| fMRLAB | Single-Trial Variability in Event-Related BOLD Signals (J. R. Duann, 2002) | 221 |
| BrainVISA | Software platform for visualization and analysis of multi-modality brain data(Y. Cointeps, 2001) | 127 |
| NIH Toolbox | Gershon RC, Cella D, Fox NA, Havlik RJ, Hendrie HC, Wagster MV. Assessment of neurological and behavioural function: the NIH Toolbox. Lancet Neurol. 2010;9(2):138-139. | 134 |

| | | |
|---|---|---|
| Nilearn | Machine learning for neuroimaging with scikit-learn (A.Abraham, 2014) | 120 |
| Humain Connectome Project | Informatics and data mining tools and strategies for the Human Connectome Project. (Marcus, D.S.,2011) | 113 |
| DPTools | Whole Brain quantitative CBF and CBV measurement using MRI blus tracking: comparison of methodologies.(Anne M. Smith, 2000) | 85 |

**Table 3.** Detailed information for most popular and conventional software.

| Features/Toolbox | SPM12 | Freesurfer | FSL | Nilearn (Nipy) |
|---|---|---|---|---|
| Base language/ software | Matlab | Stand alone | Sland alone | Python |
| Input data | Structural MRI<br>Functional MRI<br>Diffusion MRI (add ins) | Structural MRI<br>Functional MRI<br>Diffusion MRI | Structural MRI<br>Functional MRI<br>Diffusion MRI | Structural MRI<br>Functional MRI<br>Diffusion MRI (add ins) |
| Main Functions | Slice timing correction<br>Spatial normalisation<br>Segmentation<br>Canonical Variates Analysis<br>Regional Bayesian model comparison<br>Mixed-effects models<br>Bayesian model selection for group studies<br>Reparametrisation of the bilinear model | Skull stripping<br>B1 bias field correction,<br>gray-white matter segmentation<br>Reconstruction of cortical surface models<br>Labeling of regions on the cortical surface<br>Nonlinear registration<br>of the cortical surface of an individual with a stereotaxic atlas<br>Statistical analysis of group morphometry differences | model-based analysis<br>quantitative resting perfusion<br>analysis from perfusion<br>ASL FMRI<br>structural analysis<br>brain extraction<br>tissue-type segmentation<br>segmentation of subcortical structures<br>voxelwise analysis of multi-subject grey-matter density<br>longitudinal | Functional Connectivity<br>Automatic Dataset Fetching<br>Decoding<br>Multivariate decompositions<br>Image processing and resampling utilities<br>Loading and Processing files easily<br>Data Masking Utilities<br>Operating on regions<br>Mass-univariate analysis<br>Plotting brain data |

| | | | | |
|---|---|---|---|---|
| Operating system | Linux, Mac OS, Windows | Linux, Mac OS X Windows (via VirtualBox) | Linux, Mac OS, Windows Virtual Machine | Linux, Mac OS, Windows |
| Source code availability | yes | yes | yes | yes |
| Add ins | yes | yes | yes | yes |
| Support | Founded | NITRC | NITRC | NITRC |
| Statistical analysis | yes | yes | yes | yes |
| Atlases supported | Desikan-Killiany<br>Destrieux<br>Yeo<br>Volumetric atlas | Desikan-Killiany<br>Destrieux<br>Yeo | Desikan-Killiany<br>Destrieux<br>Yeo | Desikan-Killiany<br>Destrieux<br>Yeo |
| Data analysis | Matlab tools | No | No | Python tools |
| Google requests per 2017 #Science tag | 168 | 995 | 3820 | 77 |

**Table 4.** List of software/interfaces providing the pipelines for several toolboxes.

| Source | Usage | Pipeline | URL |
|---|---|---|---|
| Nipype | Software | FSL, ANTs, FreeSurfer, AFNI ,SPM, Camino, MRtrix, Slicer, MNE etc | http://nipype.readthedocs.io/en/latest/ |
| fMRIprep | Software | FSL, ANTs, FreeSurfer, AFNI. | http://fmriprep.readthedocs.io/en/latest/# |
| NDMG | Software | FSL, Dipy, Nibabel, Nilearn, Networkx, Numpy, Scipy, Scikit-Learn etc. | https://github.com/neurodata/ndmg |
| MRIQC | Image quality metrics | FSL, ANTs, AFNI. | http://mriqc.org |
| LONI | Web-based interface | Freesurfer, SPM, BrainSuit, Diffusional Toolkit etc. | http://pipeline.loni.usc.edu/ |

| | | | |
|---|---|---|---|
| OpenNeuro | Web-based interface | Freesurfer, SPM, Diffusional Toolkit etc. | https://openneuro.org/ |

### 4.1 Software reproducibly testing and claims

For some of the software tool boxes it is well known that results are sensitive operating system and the program version installed [29]–[31]. Therefore, is crucial to check the release of the program, because the software failure in the step of feature generation will lead to discarding all the research results.

To get the reliable and reproducible data software users asked to:

- Keep updated with the latest version of the software;
- Conduct all the research on one machines, or within the machines cluster with similar architectures.

## 5 Conclusion

Machine learning approaches have a potential to fundamentally alter medical image analysis via development diagnostic and clinical prognostic prediction tools, but it requires the specific approach to the data processing.

In this paper we observed the peculiarities of work with the MRI imagery software, as the most widely spread method for interpretable feature generation for further ML analysis.

We provide a general review on the most available toolboxes for the magnetic resonance imagery analysis; common MRI data types, properties and modalities.

There is a growing number of software packages for MRI data analysis representing general or more specialized functionality. We did a comparison of existed software, updated the results of previously published surveys (see Table 1).

The worldwide research are mostly done by the free available methods, because it allows overall comparison, reproducing and direct knowledge transfer. In Table 2, we summarized the number of references for open-source software, which showed conventional suits for processing and analyzing human brain MRI data, including SPM, FreeSurfer, FSL, ANTs, AFNI etc.

The summary on the number of these software properties are represented in Table 3. Also, recently appeared resources for the data processing pipelines, such as OpenNeuro and NiPype, allowing usage and comparison of several mentioned before toolboxes on a cite, were listed in a Table 4.

We mentioned questions of MRI research reporting and dataset homogeneity, results verification and reproducibility; gave our perspective view on the features, obtained from the MRI data processing, their informative content and discussed common machine learning classification problems in working with the MRI data.

**Acknowledgments**. This work was supported by Skoltech "Biomedical Initiative Program".

## References

[1] M. Behroozi and M. R. Daliri, "Software tools for the analysis of functional magnetic resonance imaging," *Basic Clin. Neurosci.*, vol. 3, no. 5, pp. 71–83, 2012.

[2] H. Ammari, "An introduction to mathematics of emerging biomedical imaging," *Mathématiques et Applications*, vol. 62. Berlin: Springer, 2008.

[3] J. S. Duncan, "Neuroimaging of the epilepsies," *Riv. di Neuroradiol.*, vol. 17, no. 3, pp. 446–463, 2004.

[4] M. de Bruijne, "Machine learning approaches in medical image analysis: From detection to diagnosis," *Med. Image Anal.*, vol. 33, pp. 94–97, 2016.

[5] M. D. Kohli, R. M. Summers, and J. R. Geis, "Medical Image Data and Datasets in the Era of Machine Learning—Whitepaper from the 2016 C-MIMI Meeting Dataset Session," *J. Digit. Imaging*, vol. 30, no. 4, pp. 392–399, 2017.

[6] F. Alfaro-Almagro *et al.*, "Image processing and Quality Control for the first 10,000 brain imaging datasets from UK Biobank," *Neuroimage*, vol. 166, no. October 2017, pp. 400–424, 2018.

[7] E. Kaza, U. Klose, and M. Lotze, "Comparison of a 32-channel with a 12-channel head coil: Are there relevant improvements for functional imaging?," *J. Magn. Reson. Imaging*, vol. 34, no. 1, pp. 173–183, 2011.

[8] J. Neafie, "= 1 . 5 Y," vol. 479, no. 7373, pp. 1–14, 2015.

[9] H. C. Moon, H.-M. Baek, and Y. S. Park, "Comparison of 3 and 7 Tesla Magnetic Resonance Imaging of Obstructive Hydrocephalus Caused by Tectal Glioma.," *Brain tumor Res. Treat.*, vol. 4, no. 2, pp. 150–154, 2016.

[10] A. G. Van Der Kolk, J. Hendrikse, J. J. M. Zwanenburg, F. Visser, and P. R. Luijten, "Clinical applications of 7 T MRI in the brain," *Eur. J. Radiol.*, vol. 82, no. 5, pp. 708–718, 2013.

[11] R. Wood, K. Bassett, Foerster, C. Spry, and L. Tong, "1.5 Tesla Magnetic Resonance Imaging Scanners Compared With 3.0 Tesla Magnetic Resonance Imaging Scanners: Systematic Review of Clinical Effectiveness.," *CADTH Technol. Overv.*, vol. 2, no. 2, p. e2201, 2012.

[12] W. A. Mehan *et al.*, "Optimal brain MRI protocol for new neurological complaint," *PLoS One*, vol. 9, no. 10, 2014.

[13] M. Bilello *et al.*, "Population-based MRI atlases of spatial distribution are specific to patient and tumor characteristics in glioblastoma," *NeuroImage Clin.*, vol. 12, pp. 34–40, 2016.

[14] S. Achard and E. Bullmore, "Efficiency and cost of economical brain functional networks," *PLoS Comput. Biol.*, vol. 3, no. 2, pp. 0174–0183, 2007.

[15] S. Klamer *et al.*, "Multimodal effective connectivity analysis reveals seizure focus and propagation in musicogenic epilepsy," *Neuroimage*, vol. 113, pp. 70–77, 2015.

[16] T. S. Kellermann *et al.*, "Prediction of specific depressive symptom clusters in youth with epilepsy : The NDDI-E-Y versus Neuro-QOL SF," pp. 1–10, 2017.

[17] B. Hosseinifard, M. H. Moradi, and R. Rostami, "Classifying depression patients and normal subjects using machine learning techniques and nonlinear features from EEG signal," *Comput. Methods Programs Biomed.*, vol. 109, no. 3, pp. 339–345, 2013.

[18] P. W. Mirowski, Y. LeCun, D. Madhavan, and R. Kuzniecky, "Comparing SVM and

convolutional networks for epileptic seizure prediction from intracranial EEG," *Proc. 2008 IEEE Work. Mach. Learn. Signal Process. MLSP 2008*, pp. 244–249, 2008.

[19] M. J. Patel, A. Khalaf, and H. J. Aizenstein, "Studying depression using imaging and machine learning methods," *NeuroImage Clin.*, vol. 10, pp. 115–123, 2016.

[20] T. Pekkala *et al.*, "Development of a late-life dementia prediction index with supervised machine learning in the population-based CAIDE study," *J. Alzheimer's Dis.*, vol. 55, no. 3, pp. 1055–1067, 2017.

[21] M. Khondoker, R. Dobson, C. Skirrow, A. Simmons, and D. Stahl, "A comparison of machine learning methods for classification using simulation with multiple real data examples from mental health studies," *Stat. Methods Med. Res.*, vol. 25, no. 5, pp. 1804–1823, 2016.

[22] J.-G. Lee *et al.*, "Deep Learning in Medical Imaging: General Overview," *Korean J. Radiol.*, vol. 18, no. 4, p. 570, 2017.

[23] P. Thodoroff, J. Pineau, and A. Lim, "Learning Robust Features using Deep Learning for Automatic Seizure Detection," vol. 56, 2016.

[24] S. Korolev, A. Safiullin, M. Belyaev, and Y. Dodonova, "Residual and plain convolutional neural networks for 3D brain MRI classification," *Proc. - Int. Symp. Biomed. Imaging*, pp. 835–838, 2017.

[25] M. Park and W.-J. Moon, "Structural MR Imaging in the Diagnosis of Alzheimer's Disease and Other Neurodegenerative Dementia: Current Imaging Approach and Future Perspectives," *Korean J. Radiol.*, vol. 17, no. 6, p. 827, 2016.

[26] A. D. Dekker, J. Fortea, R. Blesa, and P. P. De Deyn, "Cerebrospinal fluid biomarkers for Alzheimer's disease in Down syndrome," *Alzheimer's Dement. Diagnosis, Assess. Dis. Monit.*, vol. 8, pp. 1–10, 2017.

[27] D. Kim *et al.*, "Knowledge-driven binning approach for rare variant association analysis: application to neuroimaging biomarkers in Alzheimer's disease," *BMC Med. Inform. Decis. Mak.*, vol. 17, no. S1, p. 61, 2017.

[28] E. Imabayashi *et al.*, "Comparison between brain CT and MRI for voxel-based morphometry of alzheimer's disease," *Brain Behav.*, vol. 3, no. 4, pp. 487–493, 2013.

[29] T. Glatard *et al.*, "Reproducibility of neuroimaging analyses across operating systems," *Front. Neuroinform.*, vol. 9, no. April, pp. 1–14, 2015.

[30] C. Velasco-Annis, A. Akhondi-Asl, A. Stamm, and S. K. Warfield, "Reproducibility of Brain MRI Segmentation Algorithms: Empirical Comparison of Local MAP PSTAPLE, FreeSurfer, and FSL-FIRST," *J. Neuroimaging*, 2017.

[31] D. L. Tudorascu *et al.*, "Reproducibility and bias in healthy brain segmentation: Comparison of two popular neuroimaging platforms," *Front. Neurosci.*, vol. 10, no. NOV, pp. 1–8, 2016.

# Author Index

## Journals by ibai-publishing

The journals are free on-line journals but having in parallel hardcopies of the journals. The free on-line access to the content of the paper should ensure fast and easy access to new research developments for researchers all over the world. The hardcopy of the journal can be purchased by individuals, companies, and libraries.

### Transactions on Machine Learning and Data Mining
(ISSN: 1865-6781)

The International Journal "Transactions on Machine Learning and Data Mining" is a periodical appearing twice a year. The journal focuses on novel theoretical work for particular topics in Data Mining and applications on Data Mining.

Net Price (per issue): EURO 100
Germany (per issue): EURO 107 (incl. 7% VAT)

Submission for the journal should be send to:
info@ibai-publishing.org

For more information visited: www.ibai-publishing.org/journal/mldm/about.html

### Transactions on Case-Based Reasoning
(ISSN:1867-366X)

The International Journal "Transactions on Case-Based Reasoning" is a periodical appearing once a year.

Net Price (per issue): EURO 100
Germany (per issue): EURO 107 (incl. 7% VAT)

Submission for the journal should be send to:
info@ibai-publishing.org

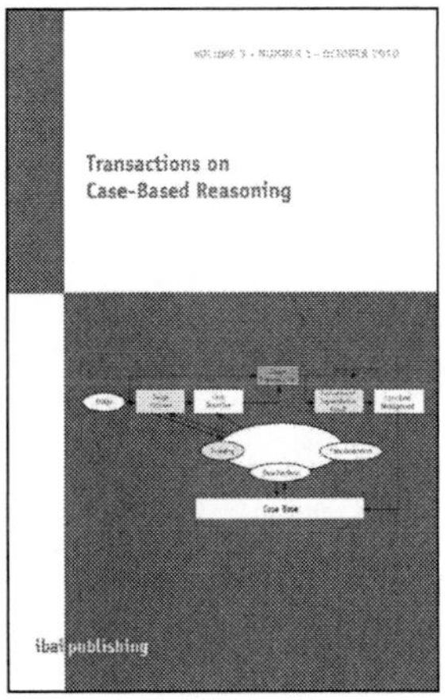


For more information visited: www.ibai-publishing.org/journal/cbr/about.html